Basic Conception Antistar Wars Program and Elements of Space

By

Kevin Kondol

Published by New Generation Publishing in 2020

Copyright © Kevin Kondol 2020

First Edition

© Copyright by author
United States Copyright Office
Register in library of congress.
1998
Basic conception antistar wars program
and elements of space

The author asserts the moral right under the Copyright, Designs and Patents Act 1988 to be identified as the author of this work.

All Rights reserved. No part of this publication may be reproduced, stored in a retrieval system or transmitted, in any form or by any means without the prior consent of the author, nor be otherwise circulated in any form of binding or cover other than that which it is published and without a similar condition being imposed on the subsequent purchaser.

ISBN
　　　　Paperback　　978-1-80031-831-1
　　　　Ebook　　　　978-1-80031-830-4

www.newgeneration-publishing.com

10. 10. 1988

Elements of space being farther stage of development even this part of nature have limits and kind of action depend from former their elements and occurence and by farther permanent contact with other elements and occurence even if such contact not always can be notice though exist.

To exploit these elements could have wide range and meaning but first it is depend from other element or elements which exploit them. If will be better this exploit then can bring more profit. Destructive action bring often destructive effect though not always as it also depent from other elements which occasionally this destructive action can neutralise and sometime but rarely turn to positive side for themself or for other elements. That is why in higher stage of development these elements which create new group and new strategic actions of one country state agaunst other, which have in tself whole range value, details, experience and possibilty such actions to turn it to most positive effects, mostly for themself. In this strategy action is apply lot of decisions and action or simply application whole mass of elements in their many range of action and by mental process to try even factors and elements from behind known system. Strategy of action though is most often decisive factor cant be recognise as only way in this action as often other play role here also like technological means and even will yet other.

Technology means despite as look crucial role dont mean that much without proper strategy which in turn come from logical thinking and next mental process. These process have decisive role but because development other kind of elements, including technological means, are understimate.

They after plan strategy will set in proper place in space-weapons, to replace it if necessary so kind of logical games, or more and application many physical occurence which come from these weapons. As for instance increase large pressure all kind of weapons on some point or points yet few time shot beams in other target if power of weapon, generator will be not enough to destroy object with one or second shot. Beside precision very important is also speed especially if come to speed of coming missile, object which have to be destroy as soosn as possible before come to target.

In such case shot of beam from laser will be cut to part of second in which such time have to be shot beam few time. Governmments and countries more civilised will mostly build up and develop defensive weapons serve for defense or in opposite case that is development offensive weapons could be notice lessen step of this civilisation. Sometime is difficult to see difference between defensive and offensive weapons but with permanent main such development these differnce will be notice.

Main effort should be to secure country citizens best standard of living and farther development with this living standart improvement. Millitary economy is only reflection general country economy development and for longer period base mostly on millitary might without care about other economy fields and citizen country living will not go to far with that as there are millitary base and if these base are frail and less working so millitary economy sooner or later will notice it when applying half measure to keep it going will do no much good as can be seen on example history mights which fall because slow down economy and culture. Short millitary success have been in countries which usually dont have these base or to small dont count take millitary task over country possibility. In farther stage of development wars eventually will disappear not as only primitive way to solve problems but very dangerous too and not necessary. So far human mind do not mature to solve this problem and try to learn from worst side . To blame everything on governments is no at total right as these individuals which are in governments grow up from rest society which keep them and educate. Here is need effort most of humanity not only few individuals or coutries. True is necessary and search for it should be not only in field of science but also in every other as this can bring often best solution. Humanity is collect from many elements but human elements are one of many on cosmic scale not biggest or smallest under many assumptions.

Only still often reign geocentric let have other view. This geocentric is no much different from what was in former centuries just earth is no consider as center of Universe just remain man as most develop and most important being. At least in opinion of many. Basic elements all beings and particles have the same. From atoms dont reach deeper to their still bigger moleculs mass which proper quality make for next new body. Cells in living organism to gradually still new ones make evetually new body. So these basic particles have been the same just their farther change, development cause rise new elements and different form of action. In case of humanity come at the end to higher form;thinking- as result these process and still this process continue to give base on next highe units of elements and process. Some organism call living and other bodies, objects by different arrangement and farther development basic elements show other range of activity but this other not always mean more primitive but simply sometime different. To little at all knowledge of laws and to little discover space range can make often other impression. By many combime basic elements these known can be change in humans these process call thinking. Could have still for instance thinking process but this thinking can be different from this known at present and other effects by this could be than from this known thinking process. Yet is possible by change elements create different process than thinking, lower or higher including with effect what for remain rest will be new phenomenon and unknown.

In cosmic scale such occurences are plenty and universe itself should be consider as one of many objects in three dimensional space. In already known part here it show that all objects from smalest to biggest have limited size and limited period of lasting. Universe also is not exception. Calculation still farther space will show her bending as one of many proof closed or limited cosmic space, because of forces, beside existing objects on smaller scale. Earth when it will be measure then bending of space will show up too. Beside elements on earth itself bending of space wil show up because of earth shape. Different calculations of earth would show up if would flat shape for example. Similar if Universe would have flat shape so on general scale such bending space as it show would not have despite objects existing in it. Because influence of forces which make this. So all it development and laws existing here should be consider under still existing and increasing yet bigger parts of which Universe is one of many. Even reaching still deeper to it smaller space, like known elementary particles and still deeper, so even if seems that it will be parts smaller than cosmic space- but in scale of part third dimension of which Universe is such smaller space not only will be not smaller than cosmic space, but still will be part of third dimensional space and real smaller spce parts will be not known here. For example again can be taken earth.

Even going hundreds km. Inside earth, when considering this only as whole known world, so then discovering later Universe it will show up that such deep space inside earth dont have much meaning aagainst Universe. Or can be this also show on example of man. If some small like particle being inside would think that such parts like neurons are basic elements and action of neurons and neurons itself are fundamental elements only known, but then still far would be to discover other human parts either bigger or more yet fundamental. That is whole shape of man and much smaller particles, molecules, atoms and part of atoms. Similar is with Universe. If known elements and occurences will be consider as basical and whole Universe as everything and nothing more, so it would be like these small beings inside man considering above mention comparison. In surrounding three dimensional space Universe is one of many objects and is subject to laws of this part three dimensional space. That three dimensional part have space much deeper and smaller that was is in this Universe yet this Universe can look like against such part like planet may look like against Cosmos itself. Different phenomenones to which Cosmos is subject in these part create elements and actions which dont come inside Universe here, or because of difference or higher characteristics of actions,even if pass by Universe running farther, or mix with other occurrences, but they are not received or discovered often. Should be not forget too about space parts, elements and occurrences on these basic more part of space which here in Universe are not included in elementary range only beside Universe scope.

In farther research in world microcosm and cosmic space should be not forget about zones there and their existing laws as it would be in comparison to look for whole Existence only on earth, in case if Universe and elementary particles, these known now, would be consider as last ones or almost last one and no more, to slow by this development. In reality whole area of Universe, not only this known, is yet more local than earth or even atom against Universe. So more easy in such picture is to show motion and action cosmic elements and all forces which are also elements or come as a result of actions unit of elements. Despite sometime some deflection whole actions the elements and forces is more determinate and concentrate on particular way. Action then gravitational and other forces and motion of elements on design track and some ways depend from higher parts which keep this cosmic and microphisics unit on describe order. This again could be compared to some part of body; heart, kidney, lungs etc. on example of neurons as most known elementary parts for living there beings. They could wonder of systematic order of neurons action in that part of body, their development and forces which can keep these particles in such order, smaller or bigger deflections dont count. However after discover whole body and still smaller basic particles in it, picture of neurons actions in this part of body will look diferent for such beings so they could comprehend orderly actions forces in this part of body. Their imposibillity of free and far reaching functions as would show up that this part is depend from function whole body when still smaller basic elements influence on such development also have too.

In some sense is the same with whole Cosmos, his smallest known elements and known all forces, dependent from bigger part in which they are included yet with smaller elementary particles or other. From this bigger part are dependent four known forces, plus other not discover yet. These four known forces despite different from each other propriety sometime, complete each other and create whole harmonize scheme still make farther whole development of Universe. In microworld atom is keep in one part, even if differenent charge, positive and negative, or still other proprietes gravitational and nuclear which each of them despite if they have similar features have often different characteristic own action result of fundamental own structure, that is other kind basic elements which on such structure build are make up, sometime other way composed. Because every element gather with smaller ones yet sub-elements, even if occasionally is make up with the same or almost the same sub-elements, can show different propriety if will be different scheme make up the same sub-elements. Similar like in sentence the same letters but different way arranged give other sense of sentence. All these existining forces come from one unit from which are depend, or from other more basic force or kind of force which can be called Gratiational force. This force is behind space field or deeper in it part, and these proprietes are different from known and unknown here forces, though some of them are the same as these forces from this former one are derived,so some similar property still retain.

This Gratiational force dont have like waves reaching action and have speed faster than light as is not depend from law in such more close system as is Universe and it influence is not only for Universe but reach also for other zones too. This force have also own smallest elements; Gratiations, but also have derivation and action from smallest there elements, particles which are in comparison to here smallest elements, particles known. But then

Gratiational force is depend from other occurrence which direct it or have on it influence and on few other occurences which in turn often have influence on cosmic ones. Atoms itself are like somehow small windows, chinks, by which is coming matter mostly in shape of waves from human point of view- from zones smaller and smaller than space in comparison to this one, when coming mostly by rotating motion in it central part, that is in atom. Should be not forget matter not only in atoms but too between particular atoms. Part of this matter and occurrences been created by radiation of atoms and their motion and collisions between them, but part is remain of the same matter from which atoms been build. Atom nucleus, his central part is more basic elementary part, especially if is count time period, like is sun in solar system. Because is difficult to imagine creation first planets and later sun, though in Universe during solar creation as result some irregular moves in motion can happen such possiblity, though it would be small such divisions of time on such long period of creation-yet sun in this period of time will be still in time of creation with central gravitation in it, despite slower creation sun elements.

On period of billions years still can happen process creation of planets so in this solar system in farther distance from sun, some next planet or small planet could been created form gas or rocks. Also in units atom creation in farther process as they are still going could show up by change of force and elements, in smaller parts, some new particle or change could happen in atom nucleus bigger than going up now.

In creation cosmic objects could be notice centrifugal activity force making too circular motion and connected by this shape of objects like for example motion atom nucleus or planet formation. So if then Universe take activity from it elementary components can be re-create his begininng and farther functions; With centrifugal force of action in three dimensional huge space or more exactly it only some part – start motion of particles in this part which circulate araund middle part of nucleus and start create outlines the Universe. Occuped zone of Universe still in formation get gradually bigger like zone of fomation of earth was in start smaller than is at present. In order next become bigger and more solid by coming greater numbers of particles. Also by still bigger Universe expansion been formed and still are new objects ; stars, planets yet other which united in bigger groups start spin araund Universe centre.

Universe itself by centrifugal force make still bigger own zone of space run the same time by three dimensional space. In his inner space formed planets, also by centrifugal force action start spin araund sun and with sun araund middlle of galaxy, so these unite in mega-galaxies, bigger groups, spin araund middle Universe. When Universe run by dark matter of three dimensional space spreading more.

Objects closer to middle circulate araund faster and longest of course than these on the edge of Universe. Like Venus have shorter distance to run araund sun that Uranus. Also during this increase own space on cost the third dimension more exactly, as increasing Universe space- these farthest objects on his edge or close to it, being under influence of spreading run most from center the Universe. These ones closer to center hold stronger gravitational force. Doppler occurrence not so always could mean permanent run galaxies from each other, yet faster if is bigger distance from each other as for last millions of years at least many galaxies should be not vissible. Motion of galaxies, their groups and other objects dont happen on ideal orbits but with some deflection so these deflections on Universe size scale, yet distance often very large between galaxies, could reach up to millions light years or more. Permanent increasing in size by Cosmos also have limit in time and space, the same like stabilize sun and planets. After stop expansion these receding farthest objects from Universe center will be finish process. Though maybe if on these objects gravitational force dont have that much influence they could run farther to three dimensional space. To know more or less age of Universe, approximate size and study division some particles or forces actions, can be calculate mass whole Universe it approximate speed and shape, which is in fact only motion objects in three dimensional space.

It was then not creation of Universe in sense of so called Bing-Bang though such theory can be modified somehow. More close comprehending rising Cosmos give picture investigation of atoms, planet, stars-as it elements -being, that is atoms, planets,stars microcosm against whole Cosmos. Microcosm is from view of human when in reality they are growing up objects for which after crossing some limit other become microworld.

    In farther much three dimensional space are also others universe with similar type of structure; stars, planets etc. They are too other kinds of universe or at all different objects. These different one other oobjects often if farther from this Universe not always have similar way of creation and structure. As other occurrences, changes make other kinds, not necessary with centrifugal active force, rotary motion and so on. These other phenomenons create type of structure inside different from third dimensional space but in general scheme in this third dimensional space are included. Previous sometime forces create often different shape of atoms, other it structure and sometime other kind than atoms. With investigation this moving Universe together with other objects araund centre, part of third dimension- with his planets, stars- should be not done as ivestigation with position only of planet on which is now living, or most, or some part of Universe, but should be look with perspective behind cosmic space so it should widen not narrow zone of such investigation. Like too with research on earth should be look on it not only with earth prospective but cosmic one also. These close model of atoms, stars, Universe have limited range because more early or later start gradually other process.

Centrifugal force or centripetal are of course in general action the same force just in reverse action.
    Notion Universe expansion can be given on example smaller objects.It is not this way as is often consider, but on example of sun where follow expansion in rotary motion so close in comparison to expansion of Universe. By creation elements and forces inside is Universe most denset than space outside so it could make assumption that there is almost no matter, like mass of sun is more denset from cosmic space. Therefore is Universe one of many objects, not whole Existence as often is suggested. Beside if then was total Existence and by this close isolate object,with law preservation energy- then would be different results motion of forces and elements than are now. Even advocates of Universe expansion have often such view and imagine Universe as whole Existence what is not only against all physics laws but also against logic, logical thinking, as if some object show expansion have space to it, as would be difficult for beings living in sun type size if they will see his expansion. Shrink and expansion may happen even after object been stabilized, but it would be on much lesser scale at least for such

physical body. In such large object like is Universe happen some local deflection against whole cosmic object with their action effect but it should be not taken as to action and structure whole Cosmos.

Waves,rays and other components coming to earth dont run on straight lines but they bend too and have different speed because of close cosmic unit being yet in motion, rotary movement bodies in this unit, itself circulation, motion of earth, yet lot of forces and elements which are on the way these waves. Light will different run on straight wall yet different on sphere being in motion, yet when light source itself will be in motion in moving space. Usually if farther from earth then bigger can be deflections or if more other physical occurrences and components with different qualities. This whole unit itself is collection of elements making sum- as is a question white light of sun passing prism and old dispute if light is whole unit make of collection colours or colours make this whole unit. The same can be question if man is body make of of many parts or these parts make whole body. Which all this depend from proper interpretation. If particular body is already as whole then is consider as whole body consisted with many elements, though as on example of earth could be notice that before it been created as a particular formed body many different elements come to make this whole earth and is already now collection these elements. Same like by mixture few colours to achieve one new, could be ask if this now is one colour make up with many or many colours created this one when process making up this colour was know. Colour itself is of course result radiation the waves and in dependence from their change can be colour change too.Should be take under consideration first cause,occurence which was before particular result happen. In this part here everything look like have particle kind of matter motion and development.

Every particle in atom have own colour and radiation of own. Collection of such particles with some degree of radiation make particular colour yet by farther connection with other particles with different vibrations, radiation- make yet other genaral colour of whole atom which divaded to smaller particles such particular colours will show. Atom then with link with other atoms and matter between atoms gradually by continous changes with new groups, still next bigger particles make up colour particular body. Parts other bodies, human, for instance eyes have own unit of elements, radiation- so that is by connection with other body or radiation other body in way for itself proper some range of result such radiation other body and it propriety will receive. If such receiving body, part will be compose again with other, in part or total, with elements of other made waves or principles own radiation, then will again mostly differently or otherwise receive colour of radiation other body, or by own propriety can such radiation repulse or dont notice. The same is with other senses of human or other species who the same bodies, their propriety sometime or often receive differently, including on occasion with another perception under shape, numbers as particular some part may not have always ability perception and division on particular individual bodies. Should be not always problem if some elements with be receive differently by different often species,bodies- as range and dissimilarity of perception was only different because these receiving elements their own shape and proprietes have.

Other proprietes and groups of body like brain or next other these differences to lesser or bigger step can overcome by next build up own picture or or comparison, search numbers elements and occurences. Antiparticles are of course the same particles of matter as are particles just with different characteristics coming from the same previous matter, when during farther evolution these own dissimilarity like charge, own field etc. developed. So antiparticles as anti-matter could be treated as lokal phenomenon because real matter and anti-matter on wide range would be destroy or been repulse during contact. These antiparticles could be of course make artificial way by radiation and bring camon particle to different mark.electrical charge, magnetic moment or all- in dependence from intended goal- proprietes which this antiparticles have on own. Somehow it have to be make sure that all proprieties or almost all been change. Such antiparticles could be exploit in wide range- in medicine to make opposite particles and charges to destroy intended bacteriums, or to prevent effects of radiation by application opposite particles and rays destroy or neutralise effect such radiation from reactor or bomb explosion. Antiparticles can be helpful to increase power of generators not only by extort emission but to create much bigger electric power and other. In creation in generator power could be help radiation itself coming from collision particles and antipaticles.

If even is not enough antiparticles so again can be by change proprietes of particles create new one or just made new one which by force of radiation after collision will change in part other groups a matter and this one after make necessary radiation give higher power for next group. It will increase by that power of generator, yet stronger and farther range sending rays, bunch of rays with change somehow it structure. Also power could be increase by making still bigger pressure between particles by action gravitational forces to cause yet bigger mass despite smaller size, or by still faster rotation whole mass of particles.

Much bigger meaning and range than application antiparticles will have discovery and application Movitons. They are smallest particles of motion -like gravitons, smallest particles of graviton. Motion as physical occurence come out in areas much smaller than zone of field here where before it was there or come to contact with elements of matter -even different from this known here, so was not yet occurence of motion there. Energy and known mass is result of motion action in this zone here as is discover so far. To imagine coming from stationary state by contact with particles of Movitons can be apply example on this scale or some comparison. Whole mass of ice would represent for example matter in stationary state. By connection with other kind of matter like for instance sun rays after a while would come to melt of ice it change and gradual coming out new components and occurences creating next and next new one under pressure sun radiation which represent other kind of matter.

If such radiation would be end whole this matter will back to stationary state though with new other components. Stationary state of ice is only for example because it is not of course stationary state. The same exactly happen in comparison in these smaller zones of space where stationary against this zone space elementary particles come to contact with particles of movitons. If such action these particles will finish matter would back to stationary state or close such state. This pressure motion itself was in turn cause by other occurrence than was motion i and then yet stiill much smaller gradually and different than motion. These other occurrences, or smaller than motion or bigger are parts only different kind of matter and energy as is known here and other derived phenomenons come out as result activity, changes contact there. So doing farther research cant be left these facts out as it will come back to it sooner or later. Change structure,situation these particles Movitons will let move objects on other principles of motion from these known. To some would be surprise to see objects moving on other principles than motion. Also by arrangement use many forces could be make occurence motion on other priciples but Movitons are basic elements of motion and by change their structure or by introduce other particles than Movitons, that is from other occurences than motion would be possible choosen object pass by space on total different principles than motion.

Sometime it could be difficult because elements this space move on pricinciples of motion and forces may try to hold object moving on other principles, but such problem can be overcome with farther development, including by application part of space field to new action. Applying Movetons can have yet more wider range including with combine elementary particles, creation by that new laws of energy,forces or other kind than energy is. Important thing is to search for such components instead of wait until next generation will start doing it. Also by this have to be reject old assumption that energy cant be created because it would be no existing then as would be not created. Occurrences, elements in this part of Existence or other are derived from next or minor which been before them- so created them. And if energy been created so man can create it also. By removing of Movitons and former yet particles would be no energy this known, or more exactly motion. Arrangement with other components than Movitons though not in this space field as have to be find in different conditions – will let create something else than energy because these other components dont have energy but run on different occurrences and principles, yet still mass of matter could be possible create. When reach for begininng of energy it will come in the end to element from which energy come like other too elements and occurences. So like here smallest known element for comparison. Important is to go in directed right way with less effort in it.

For example some animal in cage can spend lot of energy and time trying to get out, when it would be enough to lift hook to get out of cage with much less loss energy and time for it. Whole species of animals may not do this when human or one of few individual of some species would do that. Sometime few animal of the same species can do this thing, but other even it will show it to them still will be toss themself in cage other ways looking for in vain as their step of development even such task or way to get out would not comprehend. Therefore just from field of science come itself conclusion that to diminish waste of energy when pass body from one state to another have to be apply smallest number most proper directions and all kind energy and forces.

Whole space not only cosmic but also three dimensional, smaller or bigger is collection particular elements and forces which make them up. Some kind of state gather with often different from each other with unlike yet properties these elements. If such collection as a whole will be arrange different way so will come out change of unit and type of space. But yet if this whole state will get lesser then will get smaller too space and kind forces there as the same amount of elements in smaller space may cause unlike occurrences too. This whole state if will be divaded on particular elements, then eventually will come to one of this elements- cosmic space. This state gather also with many sub-units that is zones and dimensions lesser than three dimensional or smaller gradually form whole space.

In turn this whole state is sub-unit, more exactly one of many sub-unit, to the bigger part or some bigger yet state. In such sub-unit are included also other space dimensions. Bigger state have in it collection next elements but is of course bigger than all space dimensions, so by that bigger than space itself, and then gradually still bigger. If will be add to such state some value then it will become state with higher value than was before. It could be taken from such state one of elements, like one space dimension what will not cause to lessen value of state just lose of one element. But can be left in such state one of elements but take of value from it. Then such elemment will be but without or with less value. As value is particular characteristic so when whole state have it so it have too sub-units and elements of state. To add one of coexisting element with distinct characteristics to other in the same whole state, for instance coexisting element time add to space, will enrich whole state with small kind activity coming from their connection. After removing one of coexisting element other will stil lbe going on own principles. Whole state will perform own fuction too. But functions perform also sub-units and elements of state. In this case somehow if for whole state function of one elements will be still function but with very small range, so function of whole state against this element will be much bigger occurrence than function comprehend on zone such element. Usually is not discover and often have like many other factors characteristics such whole state.

These co-ordinate of functions on level of particles,elements are often receive differently so their change could be important for such element yet without meaning for whole state. When state will change in higher form will come to it other elements and action functions. Such form can be whole Existence or yet higher part still described by addition new elements and factors of which assumption is not for elements of such state only, but for whole state and even higher part. The biggest for this functions and particulars of whole this system for these elements in such forms are immense small occurence with very limited range. Estimation of some state yet form is also very different. Given element can recognise and find plenty sub-elements from which are composed so is over them when in form estimation sub-elements to elements will be from notion line similar though on distance from itself. Which mean that vertical line change to horisontal on which in long distance from each other will be elements and sub-elements. The same refer to other fields and this what in elements will be receive as a quality,durable, value and even as comprehending- in forms will be describe as a small factors divaded from each other by as elements receive it; mathematicsl patterns and distance and relations these distances between them. Yet is difficult count here in mathematics patterns, relations and distance between durability and value or among value and comprehension, yet give scale and grades between them. Such similar counting could be meet in Existence itself. Estimation will be yet other against higher forms which yet higher characteristics contain and can describe differences between scale,grade and value or expression, character, yet give them proper measurements, calculations.

It should be no surprise as to it because such occurences and factors like fear, happiness, value, character, regard and many other have own scalar quantity when created from lesser phenomenons can be in proper scale measure as physical occurence the same like many other which are treated as only notion of something. Some of them come from border of other dimensions though not space one but they are not so treat here. Is more easy to comprehend for somebody if comprehending reach behind frame of experience. Principle uncertainty can be applied by some elements to some degree as relative things are base to known things, determinated and certain without regard if this have to do with small numbers or large one and to small or big system, state. Application of it is cause by step of present evolution and it type. Is already some progress from things unknown before to probable at present. They are still unknown yet but these probable now been before in range unknown. Next these unknown at present will become probable in future so then next progress. After embrace all or almost all elements, functions and other factors of whole state all things uncertain or relative in this state will be recognise as certain, determinise and in advance predicted exactly. To higher state will remain unknown or probable until discover higher form where such kind description disappear- but for lower state probability, relativity will be determinise. So these uncertain description have to be applied to some range only like many laws and theories so can play own role in such range serve as base to next one.

Local minds think mostly local so for them to try embrance farther and new areas can dimish gradually to zero. When if want discover these new far limits have to most remove local thinking even if make sometime mistakes. This apply to all fields. Sometime way and activity of action would bring more profit than biggest inventions.

Usually every country have right to own defence. In present time it show in millitary race tendency to more use of waves such like laser and other particles, protons, electrons, groups ions, still new ones. Such tendency with farther development will grow and slow against them, that is against beam waves, so missiles and similar weapons flying with slower speed will be more by this previous replaced. Countries which dont want this sooner or later if they can will such weapons apply to keep in march in such race for destruction. Applying old methods which have few thousand years by use mirrors which reflect beams send for example from satellites is one of first methods with need of improvement as will be in use more solid mirrors resistant to attack by bend surface light, polarisation rays with other structure on surface because different mirrors, bodies reflect other bodies in many ways. With optical materials which will be made for covering mirrors and they could bend light polarisation under most desired angle which let to make less solid mirrors but the same way reflect beams than mirrors with more solid structure surviving the same time rays beam, or such one which despite lesser force of stroke with effect too will reflect beam, instead mirrors for which are need bigger force of fire over and reflection.

Making such optical unit changing among other direction and concentration light beam help from other side to build generators with less power but on the same distance with similar force reflect beams. So not to difficult calculation and complicated search with waves optic and quantun one could even nuclear physics effort help at least in first stages of development until moment later discovery in these physics field. Arrangement few mirrors in proper distance from each other also increase power of fired beams, their farther range because proper direction and reflection let on next reflection beams from one mirror to next then to next so in end will reach target and which would be no possible with fired such beam from generator with less power to achieve such longer distance. Also power and farther distance can be increase by send to the same mirrors beams from few generators. Also is known that by making proper lens of mirror could be much increase power and distance fired beams what can be seen on example much minor like small magnifying glass from which with sun beams rays can be make to put on fire some material like for instance paper or other. Also could be apply few other possibilities including sending from few generators few beams rays in one point in mirror or few points, yet fire beams to few mirrors at ones but target after reflect one object. Is possible change shape of mirror make few of them or only one with proper shape with still faster rotation.

If faster speed of rotation so is faster light reflection. Here also few combine ways could be done like sending beam from few points at ones on such rotate mirror or few these rotate mirrors in direction one or few objects, if know speed of mirror, mirrors rotation with calculation angle and speed of reflection. Possibility have these devices in face the conflic- in danger, yet next possibility descruction of them will make necessity to accomodate them farther from earth, eventually on nearest celestial objects which will be not so difficult especially with progress, yet when strong ray beam will run with speed of light. With farther progress these waves will be send straight from earth to coming object or objects on earth alone what can destroy objects without help of bombs, especially nuclear one but not making up so far with range effect of radiation as such bombs are doing in dependence from their kind of use, Strenght of send beams could be increase by application power few generators at ones sending some range the beams, yet beside other way. Have been made experiment with distant star in cosmic space. First by narrow space the slit just about size of eye. After while by looking in star rays start to annoy retina of eye so by very long or just long period could be happen lost of sight. Then slit was bit increase so light of star fall on strong lens under which been scrap of paper which was been to try light up. When it not happen lenz was bring much stronger, thicker with other angle passing of light, so would bring stronger beam from star.

After some period it give some result because gradually even not much but paper start light up and then burn whole lot. With this some help would give small mirrors reflecting more light on object or lenz. After burning paper it could be use to light bigger objects. Similar trials give results from moon and common in distance lamp or it light passing by narrow space. Such occurence happen also in Cosmos but if been effect light up from rays from distant star far tens of thousand light years away from earth so by that much better result will give sun light or rays made by man including these one for increasing generator power. With making large fire that is with use very small amount energy from light of star it could be use to boil water then next by similar and other arrangements on next goals transfer thermal energy on mechanical one or yet other. Also have been done trials to increase generators power apply beside optical lenz also electronic, sound also and other too in dependence from intended goal when principles of applying them was the same or similar. Reflecting mirrors somentime was replace by other yet stronger reflect of beams than receiving them, for direction desired target. The same time with application different method usually known been possible the same way increase few times electric power. If first stage will fail have to be done test on different particles then in similar way on next one occasionally from few sources available so in end receive strong working rays beam.

Mention above test been describe general way. Though are different method too but these one also will help on save many necessary materials, more exactly than energy. Therefore much can be lessen power of nuclear bombs for make some laser waves where not proper division energy, matter by force itself could make more descruction than utilisation these working forces and materials. It also could be make to send rays beams from far cosmic distance so in the way to earth would be in satellites these rays intensified to reach desired target. During test even in small particles large part play temperature and colour as many particles, rays have often different reactions with different temperature and unlike it absorbsion and dispersion with still other colours though they have to be fit to such small particles. By optical and next mixture with particles can be made opictal illusion on radar or in range eye visibility or ear making disfiguration or their not proper for eye, radar source other than in reality. Also it could be done by sending from few sources yet next reflection from mirror send on particular target which and radar make more difficult recognize such object as radar work on lenght and vibration some waves so other one can disturb them. By sending waves from few sources to the one point or few at ones could show impression on radar appearance few objects making more difficult to localise proper one which in addtion mention before by division many waves could make illusion other really object than is itself.

By creation on surface of object anti particles material opposite to radar waves, total different reflection of waves can be make like done by device, concentrace then next disperse waves so they will pass by moving object or araund. Is possible to send waves from few points so they will be not in range visibility, hearing or on radar but in one point of contact make occurence which will come in range visibility, that is appearance in some place occurence, object without trace what it come from. Some similar possibilities could be make in radar devices to gain by this better, farther and more exact localisation object and waves. This spot and reflection waves is visible in some animal species when their eyes shine light especially in night catch very weak waves being yet build with very small crystal kind particles. Man too can build better and much better working devices.

Beside laser waves, protons and other will come in future in use sun rays which after proper power increase will be applied to some type of object run by thermal energy, thermonuclear which is in sun sending these rays. Beside in use will be rays of beams of sound so they could be increase until object destrucion. Are known examples of action and effect the sound on human senses and body sometime make deffect or stop action his parts. With many time increase properly applied sound waves, including with some kind sound lenzes would be possible destroy organism or make disturbance of function with time application.

Organisms are the same receivers like other devices, just biological not mechanical. With assemble vibrations and activity the waves between source which send them and organism is possible then force and range sending these waves increase so organism receiving them will be under their influence so can be regulate too in dependence from intended goal, that is range and step of action, their kind from which next will be depend effect such action. Beside can be applied sound waves, vibration to mechanical vehicles but is have to be known characteristic of waves and vibration such vehicle, so next by proper application sound waves, with their rays make defect to this vehicle, vehicles parts, yet by multiplicity and bigger force, speed such send rays of sound beams. Here such powerful source for sending such rays beams of sound could be nuclear bombs which by own powerful activity making sound force, yet by proper distribution of it and direction such waves could produce bigger sound.

Will come of course in future missiles with big energy mass which with contact with other object will produce energy in shape very high temperature so high that will destroy object. Looks as is near application too coming from outer space cosmic radiation which choosen and extract waves so by concetration, pressure with next ways many times make it stronger, to create bigger mass so when necessary by proper beams rays direct on choosen target or disparse as radiation with known effects.

All these mention possibilities plus yet other could be done with effect which dont destroy life on some intentional territory, just on some period could make stop others action so after realisation own goal such others activity can be continue. Even if such action dont show hight stage moral development but can be in case of conflict consider as war with culture against destrucive one though all such action show rather on refined war. With production new kind energy, rays and all it application more should be put attention on first idea creators atomistics theory. Of course is already create in way of evolution energy, forces all kind coming from elements but with such operation manly have should consider that particular body, element under influence next element may have division. Or it will be apple, bomb or any other element. So such element gradually is divaded. Part of this element some percentage, will divade under shape of rays, other under shape of gas, yet next under shape of sound waves and so on. Some one of part to form this element make in shape radiation is of course only matter, part of element which was divide and is seperate element collect with many no matter how small,even maybe not to notice always but still is body, element changing shape and characteristics. If in some experiment part of this element been divaded being in most part under sound shape so when want more to create under radiation shape then have to be original unit such way made such that could during division create more elements under radiation shape or close but less under sound waves shape or gas.

This original element is collect with sub-elements and these one each seperate made with next sub-elements. So if want this element divade on desired parts, other elements or under some particular shape – sound, radiation etc. - have to be often change part of sub-element or yet smaller sub-element in individual parts main element so that most wanted biggest amount or divaded part from element under desire shape receive. Help in this will do way of arangement sub-elements, change of temperature or speed and time of seperation from main element. This all can be applied in some sub-elements when other left untouch or part of element remove substitude them sometime by other. During division some elements touch with next one on the way and are seperate from main part to the moment when these main one also during touch with other main make change araund. If want best to use division of element have to be most precisely calculated it sub units and all function with direction which will be use during such process. It have to do with all elements micro and macro world.

Knowing these elements or their sub-elements and factors the action or sum sub – elements factors of particular unit can be with precision predict way and factor their action in this unit. Step of precision to predict depend mostly from such precision of measurement, their knowledge and knowledge elements the unit like too action of elements and forces out of unit.

Activity of such unit could be improve in few ways and one of them is action certain system, yet more or less certain system or large parts of this system apply in uncertainn elements. For instance some group of people will do some task with positive result when individuals of this group of such task would may never do it, yet finish. But if whole group a people will try to think about the same action for few people from this group or for one person so then this action may be done with success even by few people from this group. System can be entirely or in part certain as without regard with other existing other action one of them or few could done task properly. Particular elements cant be damage by chance as is now considered but it would stand this in oposite with other hypothesis that is no effect is without cause so if happen element damage so this damage happen by some cause which will make next one and in all sum this element action will cause in certain system effect which have base in cause of action uncertain elements. In other unknown units this hypothesis, law are not obligatory and elements actions dont have there always cause. Number of elements is dont must ever bring difficulties because could be count manly effects the actions particular groups of elements, lower number with increasing quality, value of elements and by find other more easy one co-ordintates equal to amount- dont mention about higher steps than amount or maybe if not more easy, but other co-ordinates so by comparison amount of elements similar goal achieve.

Witout regard of numbers the elements their meaning will be less gradually in comparison to size of unit which will be treated as total. For example man is treated as one person and others will consider him as such but not as cllection particular elements. The same refer to still higher units including with Universe and Existence.

Et present main goal to send rays beam to some target is to damage it in one place but will be still more developed technology to send rays beam on whole object to disturb or brooke forces between atoms with which object is made, by that disperse in total or part object in dependence from power of use on small atomic particles. Rays beams will act not only betwen atoms but too in atoms divade electrons cover and particles which make nucleus of atom. Because among atoms or in atoms itself are gravitaion forces which keep atom or whole group of atoms together as one body. These forces have own structure, propriety, depedence of speed, size and other propriety so by action of rays, waves, these propriety could be brooke or block. For instance if magnets attract to itself other body and meantime between them will be introduce other body with different, neutral propriety gravitation will be brooke what is depend from size or propriety this other body. Waves all kind are also such bodies, just they have too proper size and proper other peculiarity or force such peculiarity to brooke this gravitation.

Occasinally will be enough this gravitation cut for short time so forces other bodies will act then objets with his all elements will not work as at least part of these elements will seperate by this reorganise object structure. Again with goal against effec tthis body seperation have to be done few ways to save seperation. Like to make on body surface enough strong electromagnetic power which by own neutralisation or different properties could hold or reject rays beam or other body making impossible to touch object. Because such field will be one body when rays beam next so two different propriety which one by it force or speed would pass or reject other. In creation of field take part lot of forces and elements so by speed the particles their structure, proper arrangement, pressure and characteristics elements and forces can be make defense field with proprietes coming from this activities so will not let pass other attacking elements. Also by creation just among body particles (for example among atoms) quantity which can hold pressure strange waves which could connection among them brooke. Also by make thick with stronger structure force of gravitation which hold in them the same distance between particles, or too by making more constant with less flactuation change physical quantity when with one of reason for break connection by strange elements, waves will be their appearance during flactuation changes among elements of body and breaking connection by own appearance when with continuantion may not happen.

Problem may happen with creation on particular object proper cover field when rays beam will be concentrate in one point yet by own force damage in that point object. So object should have sufficient force on whole surface by application for instance anti particles in comparison to rays beam so could such rays beam reject or neutralise especially by making very high anti energy with touch which could these waves reject. Can be make too very sensitive devices similar to radar device and these will show in which part of device attact will strike also at ones cover field with strong force direct there. Though it should be devices with speed tens or thousand of second as coming beam may have speed close to light. Help will be too calculation speed, size, rotation to angle bend waves in particles or objects which by this can have breaks or their other direction change it mean coming rays – like for example happen cut and break radio connection between visible and non visible side of moon when sending waves next to earth- which this is only one of example activities in microphysics.

Angle change of body and concentration some his elements will let on early receiving coming other elements. The same like in radar angle change of waves with their concentration will help increase range receiving objects and their quality. Other help is change the temperature in object and elements. Have yet to be other notion about temperature which is real physical occurence not just kind of notion(more exactly thermal body). The same occurence as is gravitation, motion, colours, energy etc.- just yet other structure of temperature occurence can bring different opinion. Up to now not so much is discover with it structure which of course is collect with many parts and have more proprietes than is discover. Should be measure body heat the same as is measure space or electricity with still bigger scale but should come to discover and division body heat on many parts like colour which is too various or gravitation. Temperature(body heat) have also smallest particle which can be called ;Temperon - from word temperature. By discovering this particle and knowledge of structure can be by rea-rengement achieve yet smaller or bigger changes of temperature than at present and by discovery next elementary particles cause yet other occurence than this measure by temperature. Warm or cold is apply and describe up to particular body, for instance human, when this is only smaller or bigger increase, the same as material body from micro to macro world. Temperature when is seen it effect action must have own mass and propriety which have effect on other bodies. Or exactly change in body heat.

Could be describe that body temperature change come from phycal elements but the same all known and unknown occurence come to from elements including energy cause by former elements. Though all ocurrence in these conditions come from many parts and have own basic particles from which they come so this is make up from matter collect from elements different size which make many impacts so all occurences are cause by that. To other units or higher these descriptions may not apply but they dont been to much discover and even often are doubts about their existence what is often result small development of mind. The same as often are put final limit with new discovery to mark by this limit for brain development when as is show it could be still way to go forward. That is why smaller elementary particles which still often are considered as some state of indication energy- are very solid part of matter make up with yet smaller elements and forces acting there changing, rearengement these elements until they create these smallest known. When will be possible correctly increase picture of atom and slow on screen it spinning then will show up that this so called energetic whirl-nucleus is build with whole mass tiny, grainy particles being in spinning motion and under it influence changing shape. Unit these great amount particles give quality and nature to nucleus even if they are collect with may often different proprietes. Energetic force and gravitation these particles cause that nucleus have more clear (that is more with own qualities) structure closer to matter from which was build up.

Outside parts of atom where is less attraction among particles with less mass have more influence from other atoms and forces. Like for example outside part of sun are more influence by cosmic space than inside parts. These particles from which nucleus is build also are made from yet smaller elements similar like electrons which under own characteristic cover have lot numbers particles creating electron. Just as are bigger other particles then is usually more difficult these more tiny one to observe, similar if is more difficult to observe if space is still farther away. Reaching to farther zone of space on farther distance it is simple discovering bigger and bigger from themself parts. When reaching on smaller and smaller distance- reaching to smaller parts. If to divade hair,stick on still smaler parts can reach to such small zone as atoms are and their paticles when such hair or stick will be only smaller part but not elementary part between which it will be. Such hair, stick could be divade on smaller parts as far as can let technology possibility. Can be devide on such small part to which electron or atom nucleus will be size of Universe in comparison. So like Universe is bigger so much from electron or nucleus. It is this result of calculation infinity division or multiplication. In such division will show up objects, particles-like cells or atoms which here by own activities give rise to bigger one. And from time point can go like that without end in multiplication so mean to farther distance in any ways so simple is increasing or lessen in other direction, of zone, witout end.

If will be found other co-ordinates than time or space so it would be like look from one system on other and description it shape, dimension as in comparison when look from one planet on other or from other Universe on this one and describe its borders what in limit of one system is difficult or impossible. Borders between many systems as Universes or space dimensions, time dimensions or other, also are different and difficult to pass or search.

Like space borders seperate planets or two cold system divade by hot one, though these borders in case of dimensions dont need be measure in known distances and known propriety in this system- or more exactly part, yet small one of this system.

In some systems during division with not even arrangements of elements which could make yet other kind space bending than known here, can show that during this division despite meeting the same elements will possible back to the same as was in begininng. Because bending line will run gradually horisontal then again under some angle vertical way making arch and in searching for smallest particles in this system will come to biggest then to the same from which was start search. Despite then conception that this is last unit it could become evident and is so that this unit is part of other, only are not known all ways and possibilities to reach out of unit.

As is known number of elements decline with estimation whole system. Size, dimension, shape, distance are depend from amount. If numbers of elements vanish then will be not these former mention measurements. Could be tens dies, balls in many distance from each other when counting their size, shape, dimension and distance against each other but when they will be connected in one whole so these measurements will not apply. Will be measure new one whole against next one known but in large systems bigger than dimensional at the end these known whole one systems will come under different, higher units, forms where whole of something as here was valued will be gone including with its all elements. Though there will be comparisons but they will be on different measure and characteristcis than known here. Is difficult to value size, distance of some the body if there is no other. It is similar in systems lower one. So these mention known mathemathical descriptions are only in some particular range and limits. They could with breaks many time repeat again and decline in whole Existence. So consider from other systems size, amount and next known here descriptions could have different meaning, very small or dont have any. In the way later evolution part in making size, shape have many functions when motion is only one of them. The same cause could make different effect. For example dead of some person could bring to some sorrow, depresion when to others satisfaction or happiness. So direct cause make different effect.

It could be show attention of very small application huge possibilities yet little exploit to now of optical application reflection of waves, elements their concentration and too limited application energy sources.

In experiment on desert where is very strong sun radiation with use special mirror which concetrate and strongly reflect rays have been done experiment so after while of concentration was possible to light up and put on fire even large materials. Such mirror could be made yet much stronger concentrate and reflecting so what many times will increase energy source for devices transfering solar energy on electrical and other. With yet next experiment with application accustic ways where on very sensitive surface was directed and struck on it few particles size of drops or smaller was possible make sound enough for human ear to hear it. Such sound by special optic devices can be accomodate after receive or save in other device. So from this device with help accustic means was made similar experiment to drop barely few waves particles to create sound to hear for human ear. This can be make unlimited times. So with first source size smaller than drop of water or else, is chance to make almost unlimited source of energy still increasing and making it stronger- in dependence from technical or mental possiblity. Such sound is energy particular form which can be convert to other form energy for own goal. Similar with technical means with proper calculation could be such drop direct on surface which reflect it until will come in range visibility of eyes. It is therefore huge increase energy this time under shape of light and too as in former case by strongly concentrate optic devices such experiment continue.

In next experiment in close room where fall sun rays by application lenz and strong reflecting mirror could been possible in seconds by direction rays fire up some materials linen like.

But these mirrors been replace with other structure with more cristal like surface where by calculation, bending and reflection of light most crystals make from special material have characteristic to particular concentration and reflection waves from sun rays which make sound. After then strong reflection of rays could been heard sounds enough to hear by human ear, even very weak with help too of diaphragm. With application few mirrors with rays reflection which have peciularity reflect and divade different waves or just one but in varied mirror, yet by increasing tens of time ability of reflection could been possible from very far distant stars catch and increase their waves many times what help to discover particles from which stars are build and forces working there. At present applications done in astronomy are very weak employed aganst existing possibilities. Force reflecting rays, division,preservation will help discover gravitational force particular star and disturbance other forces, not only gravitational, from planets araund star if such star have them.

    Coat with multi layer cover, polished and polarisation of mirrors reflecting objects is only one process to improve reflection ability which can be increase as much as possible. Reflecting surface should have such structure and characteristic own field that reflecting even smallest particle will be energeticl way many time bigger than this coming to surface. Or electrical increase or other way so it force would be so much increase.

Main obstacle are technical capabilities as by mathematical way can be calculate concentration, bend of waves, reflect division and other necessary methods. With progress of technology can be place in outer space mirrors of which force or source reflection will be many time stronger from source sun force. It can create this unlimited energy sources and when divide them could be transfer for these necessary for man. Problem may be to make structure which could hold these high temperature though reflection itself will be help as will be possible to reflect waves on far distance. Even if may look imposible now it will happen in future and now even is possible on earth scale already to do it some. Exploatation these source reflection would let to abandon traditional energy sources, especially these dangerous for health which make pollution atmosphere – yet by division elements may help clean atmosphere. Other devices reflecting light could be use to warm some planets solar system by strong reflection. Different surface reflect different waves,, including these one coming to earth from outer space. By application of reflection surface with strong energy power can be light up large parts cosmic space where is dark now. Also by catch from this reflection surface even very weak radiation could be very much increase with help some devices. Next possibility is catch and reflect cosmic radiation or energy sources from middle of galaxy or closer to middle universe. By proper similar increasing force of reflection could be possible create energy source many times bigger than Universe have, yet others Universe in third dimension- with few such devices in outer space or with development with continuation improving reflection of particles without end.

Every this particle after reflect could be divade on many other then next on other so after next reflection and concentration make total new one. Play have here too colour of reflect surface as it is result of action many waves to lay and arrange properly on such reflection made surface. Smaller in use these receiving mirrors catch even weak waves could receive strong active mind waves then by transfer on visible waves can be seen mind waves, yet with other application of them similar as in television transfer on visible picture of which procees is going in mind to show. It will be possible with develoment technology. These occurence happen of course in nature and could be seen on small example though simple on water surface or eyes some animals, insects receiving very weak waves and often reflect them stronger. With better technology maybe done many thousand time stronger working mirrors with yet other waves receiving and reflect. It could be in future threat for individuals if other ones will be in possesion these devices so will be necessary to make next one new- also reflect, disperse and divade waves on less dengerous and safe. This method maybe apply at present gradually for defence own country. In one of test been done on surface large object, field reflect and bending coming waves. Then direct on this body strong laser beam which without protection would damage or destroy body, but in such case covering it surface with reflect made field, waves have been reject and in other case disperse.

So when will come to destrucion by rays beam from earth or satellites could whole objects be cover on surface by layers properly reflect strong and fast waves beam.Such rejecting surface can be apply to increase velocity that mean slower coming body will be many times faster reflect what will depend from characteristic and structure of mirror. When ball will stroke ground with some force so will rebound to some degree but if the same for instance will be done on moon ball will rebound many times higher because weak, less gravitational force. If will be not any gravitational force ball will go to space. Again if this centripetal force turn to centrifugal so in dependence from this increasing centrifugal force body will be in such relation that mean with still faster speed remove. Farther behind centripetal force will act other ones from next objects but if bigger directed centrifugal force then father body should go away. But from other side if faster object will come to the body with centripetal force then faster should be reflect what will be remain in dependence from characteristic and mass both bodies. Though even with slower coming object on this other could act centrifugal force so could this slow coming object reflect far away what could be depend from structure one of body or both, especially with fast rotation receiving body. In other case when araund magnet will be put little steel pellet it will be pull in, but if to the same pellet will be given still faster speed so then will be later faster from magnet run away. Dependence is therefore from relation speed to force gravitation and mass between bodies.

These phenomenon; reflect, concentration, dispersion happen too in smallest known elementary particles so this is one of case possibilities creation such big occurences on cosmic scale.

If will be transfer energy source or energy from one unit to next, seperate one or in the same unit so could be by describe ways and yet other- create force of energy much bigger than in first unit, even if is desribe in general. Every then unit if exist always have own begininng no matter what kind of shape just of short the facts and proof it can be unknown. In example for geometrical figure such as circle is difficult give first point but in reality it is there where for instance was draw first on paper so in this point where it was start draw in some way. The same in triangle is difficult give on lines first start but is in that point from which was start to draw. Similar is with earth and all objects where could be apply mathemathics notion. Again this happen in meaning of elementary descriptions where could be lack of proof as to their trace or looking for other decriptions for begininng where this begininng start other occurence, other function or other form which in frame of elementary is not included. Smaller particle or smallest sometime mean something else than elementary particle so simple belong to different or bigger unit than elementary particle.

This reflect from surface should have every kind of body as only body perfectly black absorb all waves though in this unit here should reflect some as not all force these waves making are known. In dependence what structure body have and characteristic to reflect these waves and by change of characteristic will happen different reflection the same waves or the same reflection other divaded waves which can be gradually increase.

When blow on wall from one side and put hand in opossite will possible to feel on hand part of rejected air. When will be change structure this reflect surface with concentration it could be increase couple times such reflection. In experiment was direct strong dose very cold air on surface which back much more so mean more cold air waves than it receive. So with yet next change reflect surface when will be direct mass air temperature in approximation about 250c degree so reflect air should have yet lower temperature up to even to lowest known. Such beam of very cold air could be concentrate in rays beam similar like in laser and to own goals apply. The same is with very high temperatures especially if will be add yet other kind waves increasing effect send beams. As is still not sufficient search done occurence of temperature (thermal of body) and if are now known four kind of forces so temperature could be consider as fifth one. It obvious that is come from change other forces but these four known forces also come from other. Temperature change is like change intensity of field. In future if will be better research going on source and structure forces so temperature could be included in them. If any of these known force would be change or gone temperature still could be measure but if for instance will be change temperature in Universe from smallest known elements to whole cosmic unit, including with stars, planets, introduce lower than known temperature so would happen vanishing process other forces and creation new kind of gravitation or more depend from temperature.

It would be difficult not recognise as force occurence which could in total or almost destroy other forces when these forces cant destroy this one, yet when influence of temperature on other bodies is known. The same can be say about occurence of colours at most because of dependence from senses is not much known. Despite using these forces should be not forget too about other behind this unit and elements more basic which influence on this unit have without regard what happen here.

Is also waste plenty unused energy which with progress should be exploit. Lamps, light bulb in homes and in streets, radio, television, yet other appliances send every day and night by year huge amount of energy, so waves, vibrations going and waste to space- when from other side is waste time on mining from earth. There is no unuseful energy only such one which cant be use because of stage development applied for own use, or too because scientific or technological yet other means. Such wasted energy from mention devices could be from ones install, receive and transfer to other, and in battery for instance storage so next if necessary, even with change could be use for other goals. Though is no need to install devices close to sources from which been send as whole space is full of it and may remain potential bigger source than all minerals, yet other.

Beside that, again by application of lenz, law the reflection, other devices is possible increase much every particle from these sources to create large portion energy, then again from this energy every this particle increase to the same degree many times if will be possible what would may give almost un limited energy sources, eliminate forever if is need these mineral one in use now. If will be strike for instance metal in large room with use of amplifier then with small, weak sound will be create many more amount of waves against first one. Or use law of reflection,lenz- every ray even weak from working television- can increase plenty then storage as energy source. They are in comparison to industrial waste and their exploatation and change this not only technical possibility but higher scientific one- dealing with this problem.

Because this all energy when come to basics they are only elements, particles with different propriety and size.

Yet elements can connect to whole unit, change, divade and storage with proper condition. Battery have few such or similar source mean elements. These elements may be concentration with stronger force and with bigger amount on the same space.

So by proper elements connection with giving them desired qualities even very small battery could have large amount energy sufficient to move not only cars but even bigger vehicles, epecially by very slow discharge energy that is slow seperation of elements with the same time high energy work. Every sounds, waves from some source are the same energy material like coal or uranium just shape, compounds have different so in other way should be receive and transfer to next elements. Profit from cost depend from will in this direction and acting on it. Next example of translocation particles is this one which cause to rise life on earth. Continents of earth though far away sometime from each other yet seperate have many similar species of animals. One of reason is this that water and winds move from one continent to other whole groups seeds, bacterium and these if find proper conditions will adopt and reproduce on new strange continent with farther development own characteristic to particular species. Do not mention that first continents been as one whole. If dealing with Universe scale is much bigger and more differentiate.Space is enough fill with molecules, two or many atoms made and even happen rich enough organic units. Is many possibility for life creation. When earth come to exist already in Universe many stars and planets gone being one of addition to molecules.

All these particles similar as their equivalent on earth which push by waves, air travel from continent to continent- these one come to some planet in Universe and if find again proper conditions start divide and reproduce creating new species in dependence from planet conditions. Occasionally bacterium, germs have great endurance to survive and then adopt to new conditions, higher than is consider as are known only earth bacteriums when other may have much better endurance with longer life span. Therefore earth as one of continents here or islands was and is under influence different particles so is not so great surprise that in the end life rise here. Tough probably beside cosmic radiation and physical particles life rise here independent as this usually happen. Universe have many elements for life creation old one as new one being in creation. Part in such creation have also radiation and waves with different lenght so when find very small particles stimulate them to new kind of activity yet make new propriety groups of particles. If on earth where radiation is not everywhere the same will fall particles from Universe with sometime different radiation then happen reaction as result connection these other bodies. After while may happen adoption coming new element creating new one with new propriety of which one give rise to life. But radiation and waves act can too destroy life or also other objects including artificially made by man.

In one of experiment was apply source electromagnetic waves with goal to create very strong concentrate and far reaching rays beam.

In this test was exploit principle similar to radio transmitters so not itself produce waves by generator. In radio transmitters waves reach receiver by modulation and use of lenz to transfer them on voice or picture in television. Similar here the same principle was apply with different kind of waves with modulation with use of lenz to very strong concentration waves beams directed and increase for long distance range. So was make rays beams able to pierce object far many kilometers away. Of course electrical power transmitter creating waves was also large one. So when few such transmitters will be in use which create waves and directed on the same target then power will be yet much bigger with few such transmitter connected together with goal for higher power and waves will have longer range. Also application waves, rays beams have beside next virtue if can be so call. Because transmitters sending original waves may be in one point on earth or space when receivers which transfer them could be locate in many points on earth, space,properly increasing power to use for few at ones targets. Transmitter will be sending one kind of waves to many receivers in satellites or else and these after transfer manly on light rays with increasing of power, yet with application reflection from surface few mirrors if necessary may be direct on attacking some country objects.

To produce rays with strong power is no such big problem now as is done this from long time in receivers in radio and transmitters. More power is need make for send on farther distance rays what can be done with existing araund electric power just similar principles of regulation as radio receivers is possible done on this goal when retain existing principle of regulation in transmitters. Part of these which already are could be helpful in carry on waves to farther distance. For instance increase voice power in radio receiver still can be hear in shape of sound. Some vibrations from transmitter or receiver could be fit to object to which are coming so by next increase waves from trasmitter again adopt object to intended goal. Radio receiver is also transmitter as it remit just other kind of waves. During attack for defence could be from satellites or from earth exploiting human technology, introduce disturbance in public media of attacking country. To main radio, television,phone stations or other is possible if is knowledge lenght waves these stations introduce other,strong to disturb transmission, yet have in reserve next one if first disturbance will not do everything as should. It is no difficult this with present technology as such disturbance happen during natural storm discharge or come from outer space from many sources.

These disturbance would be employ too for power stations or wire communication despite if will be send from not wire source as principles of work are the same, yet some waves can get in by isolated lines, especially if will be change their propriety. Will be need of course plan of this communication lines the country, countries so when necessary apply waves action in design points. Important is too long and strong one waves influence.

To receive these waves that is program should be no problem if will be known range of lenght waves, frequency even with commuication from wire one to not wire communication like to satellite. Everything what here is in motion is source transmitter, just to that have to be fit receiver and also modulation if have to be done. And again of course this all is only energy and in the end on base, only elements without regard how small and how far from each other with any speed. Even if in Existence are zones that is no elements but they are still in range of Existence which own element have though very different as may be consider here. Elements here have energy because in early stages of space lower occurence than motion give rise to motion itself and before yet lower one, coming in end to stationary state. In other conditions sometime different occurence from motion cause motion alone but they been smaller kind than motion though much unlike too and also own influence have from other units in existence. Occasionally these other occurences may bring stationary state on place new created elements.

Cosmic unit have limited mainly for itself laws and speed is in this unit limited depended from action existing elements with shape of energy. Forces creating motion rise of course much early and if want increase speed with known in this unit possibility will come to limited range of arrangement as number of elements and actions is in this cosmic unit limited. Bigger effect will give gradually application forces and elements from behind cosmic unit. Universe itself against farther space and third dimension whole space will look more as small denset mass despite if look emptiness here, as stars or planets are more denset against cosmic space. So neutrino particles would be there as large particles. Beside they run by very radiate, polarized space slowing them bit though depend from them and being remains of Universe creation from third dimensional space when it start this radiation. As is obvious that radiation will not start from anything so most be particles that this radiation cause. They are deep in atoms particles, among them and behind Universe. Similar start energy in very far former areas. Pressure other zones, sometime much bigger- by next influence higher than dimensions make occurence which was yet much smaller than motion or energy more exactly. Though energy yet dont been there yet at ones but by outcome of new elements with join yet other, then gradually next occurence and elements until in gradual stage come to Universe creation, or others Universe and objects in third dimension.

Universe frorm outside look as collection galaxies with other objects in it run by third dimensional space.

They are occurences other or bigger than motion, but in very distance in Existence may happen that is motion which dont have energy as come from different principles. Energy in some sense have similar origin or kind of similar development as electricity or other forces just that start much before. But yet more basic occurence are temperature (body heat) and colour especially, because if yet dont been energy with help of some special tools from oustide still would be possible to find what kind of body heat, even close to zero or what colour element have. They would not radiate but if exist have some qualities already existing, even if not receive outside. But they could have been measure inside. They are units in Existence where temperature or colour are not because action of forces make different kind of occurences, but sometime senses could receive temperature or colour under different shape when they are still there. Yet in other parts would be in reverse. If would be possible to measure and receive temperature and colour in state stationary unit where dont been yet energy, motion this would mean that action form smaller units or other will let occurence temperature and colour to make where will reach to and exist in stationary state without need motion or energy as their create propriety will outcome from other occurences without energy or motion and still on these principles withou energy, motion could have been.

In general whole; cosmic unit or third dimension space can be picture as one element collect with still smaller elements. Other dimensions including space and times are next other elements collect with gradually smaller sub-elements. Though this everything is included in gradually bigger element;Existence and still in zones where are higher forms than elements and here just can be apply to them comparisons.

Part of these elements in cosmic condition are receive by senses which are like canals which deliver information for brain. These few senses possessed by human help receive some propriety of elements yet just only in some range. If this range would be expand then humans would discover more from condition araund with extra addition instruments also. Some senses are more develop than others. Sense of smell and taste or rather this what these senses receive yet not always are treat as should be as for instance smell or taste is same result of act elements and what is receive is act waves, vibrations and other property particles and have similarity with next property; colour, electricity, temerature, sound. It is not only this chemical property of body because everything what is in motion and have energy so smell and taste is result motion these particles as are others quality. If could be made rays beam receive by eyes so can be made rays beam of smell and taste yet other their forms because own smallest particles they have also. Same like light or sound are cause by reaction going in body. Mind analise, transform and in proper way develop information given by all senses in one total form. Information give by one sense could be in some case replace and deliver by other sense so mind will have similar or the same notion about propriety other elements. Waves or their range receive by sight could be replace in few case by sense of hearing so shape, form elements or sometime colours by this sense would be receive.

It happen in some species of bat and other some animals and insects. Waves send by one component are here by other receive by different ways and transfer to particular sense so the same can estimate quality received body as other senses are doing it in different species. These few senses create in humans make support in his development. If all senses will be disconnect in humans so his development will be hold as would not receive anything from araund. In other kind development would still evolve by other organs but in this case would be not at all or almost no farther development though consciousness would be still acting as was develop already to this degree. Few senses which man possess is not much in comparison to other animals araund and against these which could yet have been develops as is possible to have even tens or more senses by nature evolution, including more important than sense of sight and hearing. Known from parapsychology few next senses like telepathy, clairvoyant this is barely start next yet to more important senses. In biological meaning it is only unit of organs which let receive other elements or their propriety and at present development in these condition few of them only been create. If again would been develop extra few of them so what is araund, exist now would show in more broad range and creation elementary particles would be receive differently as would show again that many more elements, propriety, forces take part in such creation and only forces and energy is not everything yet what cause this development. With farther next coming senses would again show that elements is not enough and these what seem empty, distant yet small space would fill up with mass unknown elements, other parts than elements and their qualities.

With some next senses would be possible to see time dimension as is seen space dimension and discover his basic elements Timetons, comparison to atoms. It would show also that distant stars send not only light and sound waves but also other, some slower some faster maybe even that light which come in sight range. Yet would show that these light waves dont run in so called empty space but on mass and lines of other bodies and waves and by them are carry on- and yet by other change and give them new propriety. Also electromagnetic waves yet gravitation are proced by elements movement which give them these quality and yet next previous elements make up the waves and forces which in turn influence on known elements inspire them to motion to make up these known forces.

Beside senses mind have too next parts to let consciously or subconsciously receive elements from araund or in biological system of man. These other parts than senses possess some plants species as their development would be no possible without receiving other factors from environment, just difference evolution from humans look as they are consider almost as common use object despite similar genetic evolution and similar few characteristics which should in them create many developed organs, including some higher than in humans. Yet some plant species could develop own senses. If itself general mind of man let discover more than only senses could so these other qualities, particles, similarity araund also in time should eventually discover.

By change range waves could be receive by one sense what is receive by other. Like some sound, wind, smell, taste are only waves act then by adjust them to next range and lenght maybe receive them by hearing as shape, colour etc. Like these waves which are in range of sight either as light or other also could be adopt in range of hearing, taste,scent. As they been in these condition create have quality which one or few of senses will be able perceive. Can be done here many ways plus these propriety elements can be perceive by senses which humans dont have or yet other organs than senses.

Plenty by this could be discover improving man development or increase.

Sight sense receive electromagnetic waves other body in some range, general in limit 380-720nm. These waves have limited speed so without regard as to bodies distance man should see them as they been time ago, even part of second and because all bodies transmitt waves man should see only picture shaped by these waves and never whole body as perceive only range different waves from this body. So look by this that body send yet other waves beside these known which in mixture with other senses man perceive them too and that bodies transmitt yet next kind energy than only waves and these mostly are not receive by any senses, maybe some but not consciously. Beside these all bodies, particles radiate other forces than only energy as matter come very long before energy. By different kind senses would be possible receive, or by farther development, that if elements send waves, energy so these elements itself are reflection yet other parts.

So this what is consider as each element is one of expression unknown yet forms of matter and is receive under shape of element because there no sufficient senses in humans to receive farther parts or forms the matter of which elements, particles are just farther stage of development, the same like in elements again colour or waves are farther stage development. Every such body perceive range other bodies under proper to itself shape, quality and in humans this function doing senses though not only them. All occurences connected with motion in this space also have influence on receptions by senses as motion almost all this occurence made. When asking what is energy it should be back again to mention before stationary state and actions lower than energy, motion. In this stationary state when if it was possible look on it from other system would be possible to see few characteristics despite even not motion or energy. If would to do by instrument check the body heat it would be as zero here but there would show some of kind different action. As would be possible this or other way find range of colour and shape even if not radiated,yet some amount particles with next few characteristics these elements. Motion and other occurence make action also different from temperature to measure, or colours but in describe former state different acting forces than motion these occurence temperature and coulours made. They made by this of course particle matter from which this occurence give proper characteristics, in this temperature. Colour, shape etc.

In turn these previous function create energy and motion. Motion which was now in existing system of matter influence so next rise gradual occurences including eventually eletricity which like energy before, start reach to new frontiers of matter. So energy like electricity later and other phenomenons come from previous act, function in matter until reach gradual new zones making by this new occurences. Result of this is so that relation energy to matter gradually come to zero if going in farther systems- that is to moment when it rise as here other force;electricity or gravitation. Just that such matter without these forces have different structure of elements and too characteristics. Other question would be hold together matter particles when was not yet energy because is difficult imagine gravitation without energy but this hold of particles is also there cause by other functions than gravitation at least by known shape. Gravitation itself is made of few forces. Again on mention before example can be picture one cell in some part human body. This cell doing own function and is place to the end more and less in the same part of body. To the moment finish function or damage cant move to other part of body where cant do other function or dont fall on parts despite is made from gradually smaller particles as other cells and bigger parts which such motion hold action forces these part and the same apply to still smaller parts of cell.

On greater cosmic scale is almost the same so it mean that gravitation is result other units and forces these units and it function is to some degree as is smaller or bigger possiblity of balance its own force. These other units made pressure with influence on this unit, that is cosmic space from basic to biggest particles and biggest body, yet power of these units cause gravitation to exist still depend from yet few other. Body of this unit may have some deflection from general made law of gravitation but it will keep in more or less stabilized order because act of particles from behind unit and their characteristics. So need to be mention again that when search for new law and elements is necessary go beside known unit either cosmic or elementary particles these known too as would finish in end possibility to combine amount in this unit and will stop without going farther. Yet opinion have to be yet different often as is now. For instance when thinking about far away behind known unit about stationary state before energy, motion was create then it must be really stationary state where is no going any change of motion -known or not and if there is have to be behind energy, motion. On different zones could be made occurences where motion is in first stage and later come next like colour, temperature or could be even in reverse in yet different zones in dependence form follow occurences each after next one.

It should be proper understanding and notion as to discover new law and occurence and proper way arrange them and apply them right way when believe in them if they been already arrange such way, what not always is so as on before mention example where Universe if often consider as one only unit or only Existence with support from other side theory about his expansion what is not only against discovered physical and mathematics laws but also against logic. Ocurrence of force or forces can be apply on small scale but on bigger scale force is ahead of motion, energy, space as can be put question what kind of forces cause these existing occurences and elements or whole known unit. It is known that something cant come to existence from nothing or otherwise would be very primitive conclusion with science which have nothing to do. So among other elements ;motion, energy, space could not have come out of nothing and because they dont exist before so they come out from previous elements, yet not discover though here some describe before. Elements alone create motion, energy later electricity, waves. On example here could be seen the same as if will be given material to car or missile so then could start motion. So element, matter make motion itself. The same was in start motion creation where next elements, matter cause this motion by previous yet functions. Also motion of thing can cause other body but it is of course the same principle as other body is other matter unit some particles. On example too of human body can be seen action more basic elements and next which rise from them.

In human body many glands, cells extract substance carried later in organism. It is farther stage from microworld where atoms yet smaller their parts create next substances like radiation, waves or other kind what is call energy. This is the same matter as is in human body if is mean as material one, just appear difference in size in case of both particles. Electron for instance is smaller than whole atom but it is one of basic and can becaue motion result yet smaller parts create than themself. Is draw here such difference that smallest particles dont have to be basic one and this process have go to yet still smaller particles. Two big bodies if touch each other or strike will make up or lose part of own elements under shape of sound or waves. This wave, new rise element will be much smaller than these two bodies but this bigger they will be basic one as from them this waves come out. In Existence also happen on much bigger scale such process and bigger elements make up whole units much smaller elements though these bigger will be more basic one. Should be remember that matter in this matter of particles create motion and other occurences and also in space without reagard how many dimensional and what kind because create whole also time dimensions and different yet ones which are other kind than space dimensions. Matter and it particles in time dimension have to be discover not under shape or notion of lenght, wide to which was accostom in this space dimension but under shape time dimension, yet use there time measure so different from space measure because under such shape matter particles are rise there.

Unlike but on occasion similar are there many physical law. By discovering one could be discover next. Acceleration for instance also will be possible in time dimension so by faster speed can be faster done the same section of way like in space dimension. Often somehow motion principle in this time dimension are different than this one. Matter by own functions and actions also other than motion or higher, create not only elements, particles, space, dimensions but yet very much more other forms. Matter reach too behind Existence to next Existences, lower forms that Exintences or much higher. On such wide range it types are very different form each other though in general all whole it still reamin matter with many kind and scales. By own functions on some zones matter create whole dimensions, including with space and times dimension which have own particles when other areas are without elements at all, even if are included in higher whole unit. However there differences not show up under shape of size and similar here comparison of these shapes and size but under other than known here. In still next and higher forms are appear units of which matter is part though on these condition here even if will be search much and far then after many changes will be back to original description of matter

These units from which matter is derived or yet different, start in really far away forms but as will later show up many of these will be still matter just with much higher and unlike degree than is known here. But this known here and it derived occurences can be by motion and different function exploit with goal to secure planet or some country territory. One of forms is send matter under rays beam to destroy target in space. Could be set satellites with generators to shoot these beams in case if necessary. Calculate speed and range these beams is possible with few satellites shoot permanent for while rays beams which by own motion will be in seconds cover and cut some space to hold attacking objects of country territory. Similar on lesser scale are use laser beams to secure some rooms from enter if not necessary. Few of them by own motion from few side cover part of room defend entering. Motion these rays is so calculated so even small object cant pass small space without find one of these rays beam. On bigger scale country space also such security is possible because bigger force and farther range these rays beams.

They can be if possible for defence use from satellites, planes or many points on earth. Also on earth they can be use as a barrier against conventional attack as set in proper distance generators such beams can send make barrier difficult to pass. Even with not to distant range can be set more of them to cover whole or most of territory.

Additional virtue could be if attacking side dont know movement and range this action. Help would in such case too mirrors reflect received beams, also part these mirrors can be use to deceive true source shooting beams. These generators is better secure from attack, localisation and destruction. Part of other mirrors may be set in farther distance if some objects or army would pass first barrier of beams. Also in country inside territory better set extra generators more to destroy rest remain objects if they will get in there. It would help this decrease human reserve for country defence direct them to useful task. Though sometime human mind would no do that but if better organize and calculate such barrier could more precisely save particular country. One basic difficulty could be far range and power these send beams but it was some mention before.

With present permanent development science and technology using manly nuclear weapons may become less important. Part such rays beams may destroy these targets when next part under shape of radiation when apply will let to destroy or neutralise harmful action and too radiation from nuclear bombs and missiles by beams radiation with opposite propriety to these radiated.

Will come to use known in nature occurences as among them temperature, colour, taste, sound or other motion principles, yet other than motion, removing part of space, application cosmic radiation and next. Application beams base on action temperature could be as effective as other and if higher temperature yet strenght of shot beams then usually better effect when come to touch with target. Such beam may have temperature few thousand degree, tens thousand or more with this that temperature rise will be direct to middle of beam when oustside layers will have temperature lower. Temperature the same like electricity have also smallest particles as everything in these conditions is base on particle structure but again have to be mention that temperature(body thermal) have to be treat as other known occurences with own structure. Will be then more easy after discovery apply temperature on choosen place to hold or destroy attacking target. As also on such territory can be apply colour or colours to degree when will dimish ability perception by extract from matter particles such propriety to very high degree then disperse on this choosen territory as also a smell, other matter propriety. Or after farther research application these propriety so they would get in other range of senses from these so far receive.

Because it could be surprise for instance if temperatue, smell, sound will reach in range of sight with suitable waves lenght or if colour, smell would reach in range of hearing with yet next possibility. Because as was mention that even they have such propriety by particular senses are not receive.

   Next yet different example by some countries is to do to make more weak other country by not millitary ways. To such countries, usually large one, come many migrants form countries so call third world as differnt countries have different steps of development, including in course of history. If with countries less developed will come and later more with time still higher number of migrants then such more developed country have better chance fall down in economyor in millitary. If to hot unit will still come air from more cold units so next such hot unit will become more cold or in reverse. That is why is inspire by some countries emigration from many less developed countries so part of this society get to few levels of country government fields and own culture and habits apply in this new country cause it fall down. Occasionally to groups of migrants are include people,specialised of which task is even slow and gradual is fall or to made weak particular country.

Though are possibilities to hold such actions but government circles are ususally not interensted with these problems and often own action they direct for exploatation of profit for themself in period of time when in power or dont do nothing about this by ignorance or incompetence. Economy or millitary strenght was done by people which dont draw knowledge from scientist or other, sometime dont have much education but have ability or knowledge in this field what will be repeat yet in future, and this happen in all field of science. That is why one person could do more than millions or billions of people in some particular field what is support anyway by fact of many thousand years civilisation history.

To pass from one point to other and especially for desription space elements or particular space are use mostly matemathics portrayal though could be other as well. This third dimensional space is collection many elements or occurences which are also elements. Occasionally will be enough change few elements to change or dimish space especially for bodies in it existing. For example if in this three dimensional space in which Universe is too action of forces change or limit own direction for particles in space such action would make change on size such dimension.

So if action of four known force in Universe and few unknown – direct own way only to go ahead so then for bodies in range of this action would rise in reality one dimension as new rise force would make imposible to back them other way that is return. Even if would be no action side forces these side dimension still would be but in these condition usually for side ways would be necessary turn araund now maybe impossible in dependence from elements and acting forces in new rise conditions as acting force push bodies only ahead. For instance if particular force would push man ahead then he would no have possiblity return or pass other ways. So would be for him one dimension, would rise one dimensional space by just itself acting force as of the one element of space. If elements rise from start or still rise under ifluence such steady force in one way then their whole development will be done also in this way without knowledge of other. Similar action of forces is in time dimension despite that is large seperate dimension as is space dimension and have other action of forces in that part which cause for elements influence by action move in one way despite existing other.Acceleration of speed in time dimension would make pass the same path in faster period as happen in this space dimension. After removing forces could be possible also back in time dimension on the same place that is in past but this no mean is have to be the same what was before with regard acting changes so is not exactly the same when back in this time dimension on the same place as everything is in motion and subject to change.

Period not always have mean period of time as may be connect with other dimensions than time. Time and space dimension could be together one in each other because difference of matter structure and difference sometime their functions. Is show by this that by itself action of force in particular way is possibility dimish space dimension and by this directions, despite that they are still in reality araund somewhere. These changes can be done not only by force action but yet application other elements but matter is here still basic as from it all movements and functions rise. Such application of forces could be done by test but need to be put attention that in more basic particles are these prime movements which influence on bigger parts have like too other possibility. If in far space zone are or will be motion opposite to such steady here which rise also and if under influence next farther forces start move to this space here then gradually for elements in these zones start more denset matter and possibility end of such unit which will look as only one and ultimate though so many forces, functions and diverse matter with it actions would have been outside.

If is consider how to limit force functions and it one direction move so is more easy to understand by this description creation lower forces or dimensions. But if these forces can dimish time or space dimension or other so is no difficult to assume that other forces block possibility to know next dimensions of space, time or next. By their discovery that is these forces and knowledge of them could be chance to discover more yet dimensions. Like four space dimension can be imagine as more open not so bound by forces and fifth dimension could be imagine for instance as meniscus concave or six one as meniscus convex. Though still have to be underline that not only forces take part in creation of dimensions. And also that for description and search of this matemathics science can do that but it can be done also by physical means as many ways can lead to one goal. Could be add that if forces can lessen dimensions so same they could in total cause to dissapear dimension, For instance space dimension. It no mean here lessen of space, yet particles because if even gradually smaller they are part of space, so space is still, but here of total dimish of space or it part, though still will be there other kind different from space as it part of Existence.Space itself this is element collect with gradually smaller particles but elements are plenty and when compare them then will be more easy to understand this one element;space with its sub-elements.

Least elementary particles and millions smallest yet not discover and Universe with bigger more parts, all is the same space, three dimensional space exactly where by density and other physical quantity rise occurences where later are consider as amount, so as seperate objects. Atom particles, planets, stars with other objects are consider as different bodies from each other or from whole space, because density, mass with quantity made when in reality from smallest particles to very large it is one element, part the same matter in which going changes these difference suppose done. If will be take one element, matter and start divide it as is in Nature only so then could be divide this element, matter on seperate parts. On time dimensions, next unknown also space dimensions and still divade yet. One of these parts will be three dimensional space. This again could be divide on smaller parts all them somehow from this matter derived. However such division in Nature is going by measures which only little part is known here. These division can be done in many ways and under many angles such as some mention. This mean if will be take under consideration that some occurences gone to particular part of element so next occurences give matter among other density, mass, next physical quantity so then could be acknowledge that in these parts of elements rise quantity, size one particles against other, distance and so on.

But when if it will be take from point whole element so this sub- unit gradually smaller elements, gradually smaller particles will be consider as whole and in dependence from proper sight on it or notion will be no notice at all any smaller parts, only change occurences in which gravitation, density, mass are one of many and which only here can bring recognision of amount, size one body against next or distance but as is show elements, particles when it was underline are part of higher form yet for other forms may not exist or be recognise and calculate differently what have to follows if these different, higher forms and occurences exist.

That is if will be discover, find, calculate these other forms and parts of matter so next amount, distance, size etc. will no come in new issue including with fallowing consequences. For example only now is impossible find few particles with very high amount possible existing or precognision path development particular element also with large amount possibility but if quantity will be longer no problem and will be replace with higher quantity then such solving will be easy to achieve. This rise of quantity etc. was show from one side only physical description when they are many more and yet higher form will show up though this physical description show one of many possibility which cause rise quantity, distance when next one is posssible to discover.

Such discovery would be one of help to remove theory of probability and replace with theory more determinate occurences in dependence range of some unit. Also better to reject assumption and hypothesis about creation new matter and new objects in this unit of Universe as it show that by action many physical forces change mass, size, amount, density and so on, the same matter. So it only pass to next shapes which can be take for new kind of matter when in reality it is all same just diverse matter. Like rise atom, sun, planet is consider as rise new objets when in real in whole diverse matter by action forces is only change in some part of space mass, density other kind of motion so by this in such part of space rise physical different forces which can be aknowledge as new rise objects when one more time is better underline that this is only change of units the same matter when after finish action forces return to more original unit the same matter. Some influence and actions are from out of this unit in such case out of Universe but reaching on more greater scale like was in former description mention it will just be seen bigger and smaller forces and functions in whole matter make borders, sections, whole units including with Existence or more yet.

It can be take for instance on small scale sea or ocean. This could be picture as homogenous liquid even without any smaller particles. So it will be homogenous unit. Now from other unit some functions can make gradually forces in this homogenous unit sea or ocean. These forces coming out of this unit could just for instance made more dense matter in one or two points in this homogenous unit and by own inside action create one or two objects. Actually it will be the same matter in one or two points acting different but this one point, element, particle will be for itself against surroundings seperate body because of unlike action. So will appear this what could be call amount from this first point and farther could be measure distance in any direction or dimension. By farther actions would show up next elements smaller or bigger so will rise quantity, yet next by own different often propriety divade, multiply, spread to give rise still new matemathics or also other descriptions. Some bigger objects by touch can break up on very small elements yet by next actions make up new one sometime bigger, or just by touch create vibration which by other already existing elements can be aknowledge as next new particles

Bylonger period would come sections in this unit, rise by that seperate sub- units or whole groups sub- units create on higher scale from the same matter of sub- unit or part sub- unit, create space dimensions, three dimensional or else, yet time dimensions and farther or next other sub- units than dimensions. Because is here whole lot of possibility yet is obvious that by action many forces and functions are rise only seperate groups sub-units, elements, sub elements and all of this by change units of forces and particles of the same matter whole unit when on smaller scales may look that are create new objects from this matter.

And if in whole unit by basic matter action back this elements to initial state it will be again homogenous unit without any elements and particles. Will be gone what is call objects and their matemathics descriptions as would be nothing to measure against each other. Only measure would be apply by comparison whole unit against many other. But this would be different measures, measures behind this unit. Is possible to find yet minor particles but too would be outside of this unit which dont belong to it. Such units are plenty in Existence alone or just in Nature so when will be describe and valued gradually next new units in Existence or even farther later outside Existence so description and comparison will be on these measures when in reality to value them properly have to be apply measures behind this unit and here unknown.

Like for instance description on Existence agaist other, yet higher part than Existence, as also element Existon base of Existence not mention even farther element Basicon. So if may be consider that they are unlike parts from each other and describe as bigger or smaller against them which is of course from picture of this unit here and to this level development and to perception adapt when in reality have to be discover always new descriptions and laws to search new systems so will appear by this that these size and distance, yet sense of elements will be in total different and in these measures will move out notions of distance, size etc. and will come new one.

Mention before itself increase number in human senses from few known to at least ten more will show in new picture and show new cause and kind matter forces which take part in new objects creation when now yet unknown and not discover. As will show new dimensions yet new directions of space and next will show up that this space alone is just element, one of elements, part of matter divade and arrange in many what is called here directions, dimensions by forces action not all yet discover, divade on objects and make resemblance that only exist space against some object when this what is consider highest empty space this is the same as biggest object- and yet this all from one element rise. But this element no matter which one is part different form as element which gradually by changes belong to the same system. And such systems is many to discover in Existence.

This one unit, minor against other, that is cosmic unit seperate by rotating movement form three dimension space rise as seperate part against this three dimensional space by action many forces and by this with more denset matter. Objects in Universe, stars, planets make this rotate movement in three dimensional space create such force in and out. Inside cosmic unit also by force action which make changes of mass and density rise by this and still do new objects like stars and planets. By next change or forces action if would mass, density in these points bring more or less uniform activity will rise more uniform matter too, at least in world name so microworld and would be not one objects against other when unit would be still here. So no happen any change, rise or loss of matter just change in particular point of unit. It have to do same with much higher units than only cosmic. For instance it maybe show on some different or lessen scale that spiral galaxy with own many arms by spinning movement and centripetal force so should change to spherical galaxy. Milky way by this is one of arm the Galaxy.

Like describe before whole this unit from atoms by stars to whole Universe is make up by rotate movement. Is not grow including with Universe because atoms or stars dont grow but by still spinning movement reach to new zones. As it can be seen on example sun creation so here araund middle Universe rotate cosmic objects by this these on most outside parts of Sun have most distant and longest way araund Sun so the same these most distant objects outside from middle Universe have longest run araund middle of Universe.

Can be picture creation of sun and increase by it mass still new space araund so it expansion in cosmic space. If whole unit of atoms show similar action then Universe in three dimensional space increase own size. Before this solar system start it have been precede first by rotate movement with biggest gravitational force where sun is now and solar system should have limited range gravitational force and next, circular way which may in his farthest outside borders still create some objects.Like earth or sun have by circular range a gravitational field and other so same should have to some limit solar system made from one spinning movement and gravitational, yet other forces. So all depend from rearengement, regroup matter as is on some parts not only of space (because matter create space). Under influence already farther action of force, matter start make change and instand creation new objects then next one not always from elementary one, so next in dependence from influence from outside and this influenced space could start create new body which give rise to third dimensional space. In other case of action make different than third dimensional space or yet next one as dimension of time. Same can rise all kind of Existence with next divisions or bigger and minor forms.

However description of it will be from known here only or mathematics known.If in whole sea of sand grains laws, notion,descriptions man will known only few of them so should not use them to whole reality as it may bring more confusion show yet more mind limit instead progress. Law of physcics are settle by next follows process. But dont have to be any wonder as to composition and order these law. First for humans they may look very elegant, powerful, and working orderly. But it is his opinion. For some these laws may look as design of perfection but for others very primitive not organise with many mistakes. Beside every action and physical function some system whatever how chaotic or perfect, every function or action will be or may be for particular body as being a law. So physical laws it is simple every action, motion, function as only exist and which will be discover or not- by particular being or body. In whole cosmic scale earth of course is not in middle Universe as so far discover space with objects show similar farthest distance everywhere so by this Universe have to be bigger yet in size.

Part of space action in some ways can be apply in millitary activity. Such objects like rockets all kind with nuclear weapons can be for security more early and gradually place in few points in earth far even from own country.

They can be place on bottom of sea, ocean, properly hide with cover of instruments to secure by destruction or finding it by opponent. Can be place not so far from sea line or any ocean island. They could have fit up devices which after receive radio signals from far center in own country or space, yet other celestial body- will start in motion rocket device to desire target, country or few countries if will be need for it. This way could be increase balance by sending rockets from few points where cant be not expected or even from closer distance Such rockets or next weapons better be place gradually in proper conditions not to make suspicion so would not be found from satellite. If after few years will become absolete or damage, necessary could be change for new one or action these weapons try in millitary exercise, also by sending from millitary base in own country radio signals to start. Effect depend from proper place, cover and of course from fast and precisive use. Similar with technological progress it could be done with devices which send beams, yet in other place and secure in many earth points and space.

Missiles may lie down many years then use when necessary and if not so disarment them. In time of conflict if will be catch on radar rockets or other objects coming to country territory then may be send to destroy them part anti ballistic missiles from ground. Part of them to have more of them in case as will be possibility of coming more later on own territory. Beside anti ballistic missiles in farther stage better technology simultaneusly can use beams of laser, ions and next one from country territory, planes, ships with goal as fast as possible to destroy incoming missiles. If possible straight from satellites with help of mirrors to increase beams power or to deceive source of beams by these mirrors. Beside of call population to hide then answer to attack by use offensive weapons. From own country direct of course ballistic missiles on most important targets in enemy country or countries. The same time attack from ships in many points of sea and oceans and use rockets mention before in many earth points or oeans and planes. Effect is depend from efficient performance such action as would be difficult for country opponent to defend if from many points at ones if fast and secondary attack will come. The same time is best to direct attack whole mass beams and rays from own country, yet other countries and points on earth, planes etc. Also strong beam waves to paralyze radio and electrical network or disturb it.

This way so in attacking country in whole if pssible; radio, television,phone network with other in use mass network, yet in power stations or devices which make them. To try also disturb activity remote control missiles so would not reach target. This disturb action waves is very important if will be proper way done on whole country territory and on missiles. Also in that period best to destroy opponent satellites if are in use against own country and then later dont count some some case can be use conventional forces with second this time mass attack against close on enemy territory targets. This is mention in general and in part when success depend from early preperation, caution, technology and efficency done act what in large part depend from individual ability yet have influence on whole action done. Best for humanity will be if this never happen but in existing condition always better be prepared. Large portion intercontinental missiles is better to be place in high mountains where are no eartquake and mountains itself could help to some measure against their desctruction from opponent side.

Beside part of hidden early devices on enemy territory could be start from distance and help to disturb radio communication and most important electrical instruments, especially if they are small devices easy to hide.

Next way is also in defence, application derivative anti particles which have of course different electrical charge, magnetic moment and so on. These anti particles by constant application physical propriety could be make artificially even from known elementary particles. Whole groups antiparticles made of course anti matter. When come to contact particles and anti particles usually they destroy each other. With whole groups anti particles could be done radiation as with standard groups particles. If radiation will be weak so will be small effect on particles by anti particles. When large particles or radiation from particles come to destruction and with very high radiation from anti particles, particles or their radiation would change to anti particles. If will be create strong source of radiation send from satellite or from earth to flying rockets with nuclear weapons then next nuclear force in cap should be neutralise or destroy in dependence from strenght and direction of radiation or anti radiation rather. Then even if missiles reach target explosion will not happen or chain reaction because destruction nuclear particles. Radiation have to be made so would pass cover surface of missile. Even if missiles reach target and will be explosion could be direct there anti particles radiation to destroy spread rise from radiation. The same can be done on nuclear reactor in case of damage on limited space. If they are strong beams radiation from particles so for defence could be made as strong beams radiation from anti particles against missiles.

Anti particles of course are also for use for many necessary and in need purpose.
    It is known that some plants and insects are immune to radiation after nuclear explosion so they have particular hormon and yet other substance destroying or neutralise this radiation. By more exact research could in end be find these necessary substance occasionally only by change few atom particles in whole chain acid and alkali, next by synthesis make for human to secure so in case of conflict could have large importance as large importance would have application anti particles and it radiation. They are of

course general but practical conceptions as each need more exact research though very profitable when come to result.

Additional help would be send from satellites or air planes small of few centimetre or some bigger in size sphere- bombs or flat very fast magnetic way attract to missiles or different objects so could fly in few minutes there which could be enough. So few such pellets or flat small discus attach mostly to missile knob with one directed power of explosion so by it can even in part destroy missile, objets. Or after sending signal to start them they can do the same. Weight here dont have to be importance as by proper matter concentration with arrangement of it is possible make bomb of pea size.

As space is only matter in this condition collect from elements which are in turn arangement of forces in particular points also in this condition here. Can be atom of one particle spread to size of Universe like Universe shrink to size of atom. Two such atoms so spread or two Universe so shrink also can have different mass in dependence from combine and arrangement particles on some space and speed of them. Gravitation even if depend from other units but on smaller scale have pricinciple action similar to this going in air condtition or vaccum cleaner.

Before was show division of matter on some parts existence or dimensions. On very little scale could be take for example empty round glass or empty room. Also for instance could be imagine that in this glass or room is homogenous matter from elementary particles to biggest so is no elements. Now if under influence some forces outside glass or room start change force arrangement in some parts of glass or room. Rise concentration matter so appear elements which start unite, divade, multiply, touch. In next parts of glass or room may start different activity and there will be no elements but other forms. Now if in parts with elements in one of action could in particular elements happen shrink their size, dimensions, concentration of matter on gradually lessen space is possible to wonder where this matter gone when in general range of glass or room it will only change shape on different kind of matter submitt still to all matter in glass or room. Even on farther zones dont have to be mass but different physical quantity, yet still on scale of glass and room it remain still the same matter with other physical propriety or mathematical description.

Such occurences is very large numbers that is why on cosmic and bigger yet scale maybe is possible achieve plenty by trying all ways in this direction. Lessen space happen if body shrink in own size- on this condition own three dimensions,what ever body it is. Zone smaller than three dimensional will appear when body shrink to zone where three dimensional space not exist any more, will gone. Best such comparison could be value from view other coexisting to space no matter how many dimensional as would shrink two dimensional or thirty three dimensional body.

Limited speed is rather known but unlimited speed not happen in thins condition, especially in cosmic space. However in this cosmic space is maybe such possibility as some way more close to unlimited speed could have some waves, particles which dont come in range known senses or even are just not receive by these known. They would be not as fast as unlimited speed but by not being bound by known forces here may have speed many time faster than light.

In case of millitary conflict speed may play one of crucial role. Either speed of efficiently done action or speed of missiles. Yet when to missiles speed will be add their amount. So maybe much difference if on particular target will fall for instance in one minute five or fifty missiles with same power of destruction. Important occasionally or maybe decidive role may play wits and proper thinking. It could be show on example retaliatory ballistic missiles. These missiles in case of attack on other country could be put for moment on hold. Instead first can be send cheap models of these missiles with similar looks like just without nuclear weapons but can be receive as real one and destroy so it will deprive this opponent country some amount anti ballistic missiles. Second time could be send cheap models too then third time could be send real one together mix with cheap models, when opponent country may deprive most antibalistic or other missiles and weapons for defence. Also could be second and third time send mixed missiles so attack country would dont know which one are real so then in end these real one to be send. These cheap, false models have to be properly made so would resemble real one also or radar. However best defence is permanent stabilised very high scientific, cultural,economical and technological country development. In dependence from size and number population if particular country will achieve such high standard of development so next one aginst it will be far behind so this will be best warrant of security as will be difficult for other such country to attack.

Some nations and large on earth countries not so much economically developed cant for instance stand up to some countries in western europe either by economy or millitary. That is why main goal every govenment should be gradual most possible general economic development with increase living conditions. It is maybe most difficult but happen be most important.

One of possiblity improving whole unit is improve deceptive elements quality so gradually should after this whole unit be improve too. For instance if in some less developed economically country would be improve most or all citizens to level of some develop country so then gradually such whole country should similar level achieve. At times need to be improve part of elements so in specific unit achieve quality of rest higher certain elements. But every element deceptive need improve so could improve also whole unit. Yet beside this to improve some unit can be introduce par of better elements from other units. Better not worst as worst will detoriate action of whole unit.

Such develop country would win war with other even without single shoot. Such war can be win economically by involvement other country in economical arangement, investitions, political commitments with that is have to be cautious because it would be exploit by other country in right moment. And is great ignorance to help such country economically, technologically as long is and remain dangerous for this first country. Most of maybe just little help to make it more depend.

Space again is nothing else just matter or part of matter shape in some form on individual space. Because everywhere in this space matter will be found or it will be among galaxies space or wall space or atom as it is the same three dimensional element. In this space can be receive many kind of signals and sounds coming in range of hearing.it Also have to be consider that there is not at all sounds without meaning only sounds which cant be understand.The same as many other things like this what is undertand as total emptiness is the same matter as most denset object. Between two planets sometime is consider that is empty cosmic space when in reality this empty space and planets is the same matter on broad range. So in whole system emptiness cant exist just matter with many state of concentration and mass in particular zones what make impression of existing many different objects very different from each other like atom, planet, human, radio- but when to level of these concentrate states, mass, and distribute even in whole system then matter will be more homogeneous. Not at all becasue mass and matter concentration is one of many reasons elements creation. So it can be draw some conclusions, suggestions as to this part of Existence.

1) All quantity bodies of whole system are directly proportional to amount of bodies whole system. (So amount characterize quantity and few other features).

2) In reality everywhere there is no units entirely close.

3) Mass of every body in motion in moving systems is never identical against each other and is depend from inside elements and forces which create them and outside acting on them.

4) All in-constant quantity make permanent quantity which are to some degree directly proportional and depend from in-constant quantity which create them.

Or; All constant quantity are to some degree directly proportional to all in-constant quantity which create them.

5) Relativity(indertermination) dont appear in whole system where is limited number of elements and occurences without regard of their amount and is know whole or sufficient numbers of cause which make particular effect.

6) If all occurence surround nature would be not determinise then free action of elements would be often unlimited.

7) One of characteristics accidendal events is their dependence from whole determinate system.

8) Accidental events are simultuneusly absolute events just their description happen by unlike action to absolute events which are next for them accidental events.

9) Cause and effect are two diffeent expression the same quantity(Or two shapes the same quantity).
$C = E \cdot C^2$

(Every cause is already new effect arising from other cause as effect is cause in less or bigger step to next effect. Even if effect sometime if much bigger than cause but is next then cause to following effect.

10) If distance all elements and occurences in whole system come to the zero then will complete disappear relations between them as a size, shape quantity,amount and other existing physical quantity.

(By this have to be consider distances not discover yet and unknown between all features of elements, like for instance relation quality to value, colour to sound).

11) Sum of elements in close system is always constant against whole system.

12) If distance between elements or elements and whole system in their last ends in system will be unlimited then cant be define relation between them for elements in this system.

13) Relations, measurements and dependence exist between all elements and physical quantity and their characteristics without regard of their dissimilarity and distance among them.

(In this are include measurement and relations such as between colour and motion, value and amount or amount and sound and colour etc. Relations and measurements not discover yet).

14) In infinity calculation there is no integer numbers. If some particular straight line or quantity grow infinite then integer numbers which describe them become gradually fraction.

15) In existing condition rest mass of every body cant be even to zero as then would not exist. As from the same reason cant be no any zero quantity of every physical feature of body as then this physical feature will disappear.

Or; If any physical feature or quantity of every body would reach zero point in these conditions then will cease exist as such feature. Zero relation then can be describe to some degree.

16) If on some limited surface, field or limited straight line can be mark infinite numbers of points appropriate gradually still smaller in realtion to this limited surface, field or limited straight line it mean then that infinity also have own limit in relation to particular quantity.

(That is why whole third dimension despite unlimited numbers elements it contain also can be limited. Relation of elements it contain is proportionally reverse of these elements to whole, that is to whole third dimension.

So is why in general for elements is no end in limited system of third dimension. To find it have to be apply other relations and measurements in this dimension or apply measurement of higher quantity or realtion for meaning other than elements).

17) If every limited quantity is bigger than amount then unlimited amount have limited range. So is finite.

For example on line with lenght 25 cm. Could be mark unlimited numbers of points. Despite that every of these point will have some lenght their numbers will be without end. So then this unlimited numbers with the same or varied lenght will have limited size 25 cm. It show here dependence and features, in such case dependence among quantity and amount. If amount would be higher than quantity reult would be different. Similar in case of third dimension. Despite unlimited numbers with different or similar size it also have limited range or size. Just as in former case unlimited calculation limit other quantity or other physical feature. So in case of third dimension about his end will decide other quantity or other physical feature or features. Similar is with time dimension or Existence and other forms where to find end in these measures have to be found other quantity and their comparison. Like here between distance and quantity, amount and distance, quantity and value or value and amount, yet their mixture or other features and quantity known and not.

    18) In limited zero quantity or close to zero where exist unlimited numbers of elements unlimited distance between them could be zero also or close to zero especially against others quantity.

By zero quantity have to be mean comparison one quantity against other as if some quantity if exist few feature already contain. If in some quantity can happen unlimited numbers of points and if distance between them also are unlimited but as close to zero if quantity itself is the same. That is why eternity or infinity in some quantity like third dimension for instance or space dimension or other yet quantity are or could be as zero quantity or close to zero in comparison to others quantity.

    19) Matter is always permanent against other occurences, features and physical quantity.

With this that understanding of matter have to be widen on still bigger scales and this what come under notion of matter that is unlike from each other substances in many place but what on general measure will be matter despite wide difference in many parts and forms. So maybe not everything change as is understand. Just laws and occurences in next part of Existence, forms are often different.

20) If at least two bodywhich lie against each other will bounce (push) third body with power even to square their distance then motion such bounce body could be unlimited.

(Dont count influence outside force and how solid structure body have, plus unlimited energy from inside or outside.

For example if two springs in one metre in distance from each other will bounce round body to each other with the same power then can do that endless if have endless energy for it from inside or outside and enough solid structure for it.

It have to be picture matter as was mention before as substance with very wide range, in many place very different and occasionally as may seen opposite to matter in own structure and this because so wide range. Though something never come from nothing including matter but somehow on such wide range this understand from nothing lose importance as different structure and actions there happen. This substance, matter if in some it part, yet other part will change or divade then rise in these parts whole many existence and other forms from existence or lower than exisence. By different changes and divisions could rise and they do whole natures or dimensions in natures, times zones and other yet what is only change of structure the same matter. Such change is rise of dimensions, states or higher existences, Then farther change in matter in this part is rise energy, motion, mass change creation many universe in third dimension. So rise amount, quantity and other characteristics though all this nothing more than change in just matter. Brain structure is also matter and going there process make abstracion notions or next one, but this is only standard physical process as every other araund just with that difference that everything is different from each other more or less, at least in this discover part of Existence, so farther up there maybe not the same.

As reign before and still happen now view backward somehow that put man over rest animal kingdom or else so same apply too for mind and rest of physical occurences consider them something lower when in reallity is all the same with this that every physical process in not exactly the same.

On earth and Universe dont reach farther rise elements and occurences when yet was not humans and will when there will be no humans, yet on scale very often much higher than humans can do so then other occurences and elements could do much more than man can do so by that they are on higher stage of development sometime than man is according to his understanding and value of development of which mind is only one of them. If man could bigger step in progress achieve then should get ride of these limited conviction especially that do most damage from all species. So this change which is now observe and other yet not discover units of change this only change in some parts of forms, matter. In some parts of matter because not everywhere changes could happen. Though to discover new elements and occurences then still higher forms have to be apply new laws or new action of logic and mathematics. If want to find begininngs or ends infinite now elements like dimensions then mathematics calculation have to find co-ordinate of infinity or higher range of activity in which infinity at least in this way will come as sub-unit of part higher than infinity. To discover higher forms in Nature or farther in Existence will be in need much more higher mathematics science than known at present.

On many forms mathematics picture will be not enough as will show parts where undertanding begininng and end will be not or at least under notion begininng and end known at present so will be no sense ask about begininng and end something if there will be different range of function. Do not mention still farther reaching. In other areas with unlike forces action dont have be there elements or amount as will be there homogenous matter, yet in other dont have to be distance, quantity but will be different physical features.

It can be add that in own action and life man look on everything manly with own point of view but no from interest other species or description different general point that is why may limit more own knowledge. On known condition change or farther division if it can be so name, matter create new particles and motion. Then motion by change particles create next occurences in this so call life and next division or change in motion in this range and change few particles, create mind then thinking. So life is one of form motion here, changing particles and so thinking call differently because some amount other elements. By change some elements in this already existing division could be create yet other process than thinking(Robots, yet when improve), life without thinking, which is known sometime. With discovering new zones still new possibilities like thinking without motion phenomenon or life on other principles than motion and still new farther possibility.

In some forms by force action matter occur in few unlike dimensions. Third dimensional space already discover, in some part, or often barely understand time dimesion which is found in the same place because of matter difference. Time dimension incorrectly comprehend as four dimension. This four dimensional space is yet not discover when time dimension is in total unlike dimension with distinct basic structure. Unlike- though in the same place dimension. If would be remove from space dimension then will be living here without time or time dimension but rest would be the same dont count effect of such removing. Also wrong calculations and understanding are happen with calculation of relation of motion to time change. If in three dimensional space will be point A and point B then with speed change will be possible pass distance between these points. Space yet still will be the same. Likewise is of course in time dimension. Distance between point A and point B could be pass in faster period and time dimension will be still the same. Three dimensional space is without end in time so is the same without end time dimension measure in other period. Therefore with change only of speed could be pass slower or faster section, distance in both dimensions. If from earth to distant planet of few years of light away somebody for instance would go travel with speed of light then back on earth so according to some theory should come back younger than these on earth in the same time.

But if only for instance such travel there and back would happen in few seconds then should be no any change notce despite more faster speed. There are not take under consideration beside change force or other change all co-ordinate and units. Could be say that for such person travelling so fast space would move faster when it was really no change just travel was much shorter so only space was pass faster. The same of course is in time dimension as similar distance can be done in few months or in few years. So simply faster when time dimension with it objects was the same. These objects up to smallest particles there comparison to atoms particles here in third dimensional space. That is why smallest particle there Timeton should be two and half time about more heavy than atom particle is here in comparison because time dimension is over twice smaller from space dimension as create here as it is only three quarter dimension in one direction as cant come back to past, so therefore to place matter on area over two and half smaller should give about two and half bigger mass, weight not to mention local difference proper for each dimension. This of course is not refer to whole dimesion. Main reason why time dimension matter was not discover is only five existing senses in humans so cant receive radiaton from particles of time dimension which send different range of waves than human senses could receive so barely is feel by other sense and general development. Reason is also lack of vision in some people with lack of imagination and indisposition to find new ways beside this what they have deliver by others and still to small general development.

To isolate or discover this particle from time dimension have to be first find matter which time dimension have or properly out of which is made, as usually is claim in science about time dimension so should come to mind that if after all if time dimension exist so should be build up from something, some different matter but if somebody would dont consider so then is no scientist if is about them at least in this case. Such matter or particles should be more in lenght shape as time dimension is raher about 70% of dimension in one way as cant come back to past and forces pushing matter still ahead in one way more such shape should it give. Radiation from such matter could be transfer to matter in this dimension after if receive or find it first with this case that this space dimension is three dimensional and time dimension not even whole dimension as particles in it. Yet beside that it is not space dimension just time dimension so objects should be measure under this angle and in distance of time no space. Whole radiation is also in some portion as in this part existence is particle structure of matter and already time dimesion, is divide on gradually smaller portions but if matter in this part will be discover so distance will be measure along motion this part time dimension but not space in here in Universe. Unlimited speed exist because if rise space and time dimesion unlimited on these measures and particular forces create them on unlimited measures undiscover yet. Just that speed of them is not included in frame of motion because dimensions have more phenomenons than motion which in this part, in universe are so far unknown.

Part of matter that is energy should be better exploit as energy source mention in general before. They are waste such energy sources as light or sound waves coming from sun and universe to earth. Is no problem now to install araund house devices exploit sun rays transfering it to electricity. To this should be apply with improved devices exploit not only sun light but waste waves from light bulb in homes and araund, street lights, television, many other and also exploit waves, vibrations of sound and too transfer them to electricity. Just watches or calulators exploiting even weak light of bulbs or televisions it is only ersatz of this what yet can be done. For instance if to avarage home will be install in few it parts inside and outside- instruments receiving light and sound waves so this is already large source of energy. Manly in day or night these instruments may exploit sun light, moon too and much more, reflecting light form lamps inside, outside home waves and sound from television also from radio, even from human speech and waves make by human activity, next from ovens receiving it waves. By portion this maybe no much but by many hours in day and night is together big energy source. These devices is possible connect to electric network and after transfer from light, sound and heat waves on electirc energy transfer to main electric network.

Many details this technology are known enough but not much in use because of greed and ignorance what make loss and delay in farther technological progress. Such devices may be connect to street lamps, buildings inside and outside, factories, ships, satellites and all available light, heat, sound energy sources- small and large. In city centre when is most usually motion and noise will be handy instruments exploiting sound waves on electric energy what even may help lower noise araund as instrument receiving concentration these waves inside them take them also from araund. For example plenty instruments on some space which snatch sun light cause on such space lessen degree of heat the ground by concentration rays inside them. These instruments exploiting sound as energy source may be install on millitary firing ground, night clubs, concert halls etc. where is their hightest energy source. In developed countries it may gradually come disposition about instalation these devices in most buildings and places where is biggest possibility their use even few of these devices or just one in dependence of use, exploite together light, sound and heat waves if possible. Is possible too of course deliver transfer energy to battery to use it later. One street lamp often with large light power by many hours to morning will deliver to device using from it just light waves or even some sound and heat waves to light later home or later this lamp again by some time what depend from devices which receive these waves and optical devices increasing them to higher degree.

By these mirrors, optic instruments, lenz every receive waves could be many time increase so by this increase many more received amount of energy, even by known ways or by improved as by application optic, lenz, law of reflection, concentration, proper materials-could be every waves increase almost without end, especially with technology in future. Smallest known portion increase for instance to size of light average light bulb, though for such instruments will be enough just few time increase this power use on wide range to very large degree. Such transfer waves use on other device like in laser. These devices will lower dependence from limited energy source like oil, coal,atomic energy, most in countries which dont have so much natural sources. Such countries will do best if will invest in develoment these technologies as with their improvement and improvement ways increasing power by optic, lenz, mirrors etc. can become almost self sufficient in this regard. Cost to install and improvement these instruments is actually not so high against traditional cost of present use existing source of energy. Itself this cost should be less or no much different from present cost of transport oil, coal, gas. Dont mention about farther cost of extraction, industry accidents, pollution, health cost and loss life itself which is many more over against these application of devices. That is why investment by country and support of construction, installation and improvement these devices should be done sooner than later if step of development mature to it as often people go in worst line to improve own living standard.

Here is somehow good, cheaper and clean step to this improvement. All kind light and sound waves could be use here even these in lessen amount in night but by longer period still energy receive large amount. In farther yet future may be use these waves and body propriety which are receive by sense of smell and taste as they can similar way be transfer to energy need by people.Source for it natural or artificial is great amount. All these source like atoms radiation may be use after energy change in install instruments to mineral production where to transfer to special chambers will happen temperature lower of them with addition of next waves with factory and laboratory source to production desire mineral. To such chambers could be introduce from receiving devices and increasing waves to transfer in shape of gas when in chamber precipitate it and next lead to it coagulate. To some degree this may help less dependence from mineral mining, mostly in production costly minerals with goal of profit this production until improvement technology. Because araund in shape of waves, sound, heat is plenty waste left as mineral sources as well from human and natural activity. Everything is source of energy just is very small step it use and here is whole Universe attainable as whole Universe is made of atoms. For few percentage of this what is spend on millitary already, gradually these devices can exploit known sources energy and radiation, including in microworld, use for production energy in need and next for production minerals and these sources are rather unlimited with progress.

If take under consideration some town, either small or many millions inhabited city is not difficult calculate profit coming from it. Devices use for receiving electromagnetic waves can be install in all or most buildings outside and inside connected to general electrical network, yet in public place, factories, steel works-every where are coming energy sources these waves. Could be after transfer to electricity send to power stations and from there to use for whole city. If in first stage of technology chance to use part of loss energy after receive it by optic and other devices then is already improvement. Plus now need be add solar energy, moon, ultraviolet waves, infra-red, cosmic radiation,heat waves and sound coming from human activity, so all this with improvement should give most return waste and natural energy. So eventually, yet in future if will be waste some amount in city- coal, oil, gas and other- so should return in time most of it saving by this minerals to high degree and with much improve ability of better receiving and trasfering by increase power would be possiblity have surplus dont mention about mineral production.

In experiment in special chamber with accustic application, lenz, been possible to hear sound from metal size of head a pin struck a metal thin sheet then increase it to size of noise for human ear. Every then small particle, wave could remain unlimited almost energy source in dependence from use method and material.

In such chamber even few, main source are accustic instruments; lenz,mirrors with no need of energy- where most weak sound was hear and in next chamber increase then increase again by sound lenz Here somehow to reproduce such high sound in accustic chamber was present device receiving sound waves and transfer them to electric current with help mirrors and lenz also. So in result that from very little first use energy source was possible receive very large one against first one in this experiment, dont count yet any propulsive energy source here not use.

These experiments are known but with their application and increase on industrial scale and use initially light waves or heat it can be achieve good result with improvement on industrial scale. So beside receive these waves could be construct whole factories specialise in this type energy production even with most known technology means. Man dont need now traditional, dirty, expensive to transport energy sources. These eventually improve factories on these describe principles even in general, should replace old one as investment will return not only in money but more imortant in health and life improvement. Next with progress molecular accustic produce more minerals.

It could create additional energy sources.

Dependence of energy strenght is from degree of use and technology, optic instruments, lenz, accustic, concentration, mirrors, waves reflection, colour, temperature and materials use to build it. At present in radio astronomy can be receive and divade for analise weak light distant stars by mirrors relection which can be improve as by better application lenz and colours because in dark space araund is better to see minor light and increase it strenght or colour. Though now is already know way to increase power from one atom to strenght of whole Universe enough to destroy it. Whatever may it look now. Now with best microscopes enlarge millions more still is possible to see very little reflection from energy tiny molecules and next sound waves. And this can be in future improve. This received reflection is possible increase and it use above mention accustic ways or optic still can such sound increase. Is no difficult calculate how much was increase power from one or few molecules to such big scale with help other energy source but could be otherwise. Yet can be still more done by change light waves on sound or reverse still increase them that their power. If it can be done from such little energy source so with application large original energy source could by these ways have big permanent energy source. If optic instruments, telescopes, microscopes increase picture many millions time more so similar could be done with increase power of rays. Whole laboratories and later factories could this way create with smaller supply large source energy, waves and these transfer to electric energy use for city or factory. This can be cheaper, clean power stations of future, optic, accustic, this kind energy supplying and more profitable.

Such plant for instance with receive sound instruments from outside may be construct close to some noise source, sound waves. This instrument would receive sound waves with application lenz and other optic instruments concentrate intensity waves on gradually smaller space increase this way their strenght. Then after pass to special accustic chamber again increase strenght with help of waves reflection with next few application so in end by cell change to electric energy. With few these instruments sound waves receive and transfer by day and night would be chance to get lot enough energy for electricity production. And sound sources is no short of supply araund, most in big city. It is in general describe way without more details but could be apply better yet and different in dependence from kind of waves, primary energy as it would be also heat, light waves or these tranfer to sound waves and too from expected strenght, though here power have to be create biggest possible if want replace traditional energy sources. So development and improvement accustic or molecular accustic, lenz etc. will help with this achievement what is always profitable because with great improvement will be gone problem about lack of energy as these would be without end and dont need large cost spending with exception their construction and improvement. So with this or other means should be start even minor mineral production as these many waves araund have mineral chemical propriety. It could be add about production that with present technology is possible start even now construction on greater scale especially that accustic instruments, optic and similar are main energy source

Radiation waves itself in greater yet step is possible to use in radiotheraphy most in case where is difficult to heal by standard medicine and also to save people with radioactive sickness after some damage of nuclear reactors and other radioactive source. Devices sending rays could be use in special chambers with more amount group of people and in such chamber radiate waves remove or neutralise radiactive sickness. Many waves have chemical structure to do such task and extract from proper minerals and ozone gas also could be help in some case. For damaging radioactive element could help artificially made antiparticles in frame present possibility as opposed to particles so with still longer acting antiparticles should destroy radiactive particles in body or soil. So by application radiation, radiotheraphy in this chamber save from radioactive sickness. Some minerals addition may help in few case such as selenium, cobalt,hafnium. To this goal may be add interferon, leucocytes,properdines,lexicon with way of body immunisation when better stronger body working have more chance destroy damaging factors. Three pair atoms oxygen make ozon which in atmosphere absorb few damaging rays for humans. Atmosphere alone more absorb damaging rays from outer space than was send by people to atmosphere so by creation in chambers radiation from these elements could be absorb, destroy radioactive waves as done in atmosphere, with few change.

Though again of course in proper dose without damage for body. If these elements in atmosphere save earth from damage activity they could do this in body with proper changes.

Accustic action is known in nature like in forest or among rocks in mountains and it was use from centuries also in castle for farther or stronger voice increase. It depend too from instruments sending these waves and reduce waste energy in final instrument application. If in choosen room two high sound metal sheet strike each other or water drop on this metal sheet then is already create some amount energy when energy source is basic in metal sheet as is in room to make this accustic echo. So main energy source is in reflecting matterials when creation first sound waves could be acually only ones.

Investment in it can be profitable as with one just room will be send later back large source of energy being eventually permanent source of it. In the end better if will come to such undertake because short of traditional sources and grow technological progress. Here is possible change or use light waves, sound, heat with application parabola mirrors reflecting up to 100% waves or in last stage change them on necessary to own use. Beside optic instruments to help for increase, concentration are materials reflecting few ways and these waves are different also as depend mostly from materials. So in last stage under shape laser beam or gas beam from laser- these waves as was mention can be in special rooms precipitate to coagulate made at least necessary minerals even in small portions as the same matter can be under many shapes dependent from reflecting materials so from small first source mineral would be made in pure structure. Similar methods is chance to use for increase strenght of laser or to send strong beam on farther distance. From other side devices which almost at all absorb waves could be use in test nuclear explosion with large strenght of sound so some waves will be absorb when send more under light waves or other necessary in dependence from accustic materials.

Many directions are to improve economic state yet still better and faster development. These which help in progress ordinarily get least out of it. So here is one of main point in farther progres as with goal to achieve most in some field of science and economy it could by gain individuals with ability offer them high reward if they will reach in need intended goal in some field, or in medicine, economy or science or with yet economy, political reforms. It will be good profit to spend for it millions to receive in end billions from it, plus next; life improvement and country bigger strenght. It not happen to often in history despite possiblity large profit many kind. But are spend large amount instead for individuals or occupations which bring no much in comparison or at all to human progress or better living standard, especially in comparison to mention before individuals. This show yet no to much advance in some field human development. If country will make politic of reward for these citizens which bring progress for humanity so just itself make of this political decision may bring more profit with time than many inventions. Often now are spend large sum on reasearch done by people who dont have to much ability wasting time and money for nothing or for no necessary research.

Example country defence yet to destroy other would be politic or way lead by choosen country activity to interference and payment for people in opponent country in intended own goal. It may happen by support individuals to be choose for important position by media support yet next ways. So to support this individual or few of them with knowledge they will do enterprise with line need for own country. But it is possible to support vey incompetent without much ability person, more even with knowledge that by doing this or other decision in politic or economy this, these individuals will make country more weak or bring to disaster. If such person will be remove support next one as much incompetent who yet more will put country down. Even if here will be no any millitary action but this way is possible yet more destroy country than by millitary action. Beside it can be support intermediary individuals in opponent country employ in few fields economy, banks. It may take longer here with cautious action but this country yet few if necessary may be much weak after time with lower level in many fields or general as it will no longer be to competive for own country. Sometime for different countries because of system difference or culture different method yet unlike have to be apply to achieve similar goal. As in every country are people that for them country right dont have value or because personal games they can be use. Mention too before way with help of migration to opponent country can be use as well.

To these numbers legal or illegal migrants maybe add people and support them financially if they take important positions in economy or country politics make by activity wrong decision to ruin country or just part of it. Beside add maybe people with virus infections direct them among population so can infect large number degradate society increase though dont necessary it have be virus to bring end of life. Is possibility by genetic program change this virus so it will act on particular part of society. They are countries or social groups more or less immune to virus and parts where some type of disease occur and other place not. As for example sclerosis miltiplex so often happen in tropical regions. Dont need to add that such introduce disease or virus in choosen country may do more damage than war. After war nation could fast rebuild when in this case scheme virus and gradual with caution done country degeneration even for some time may back down country for many years back depend what method will be employ. Some way is more easy this to to if country is far away from own with different occasionally conditions to make it more safe for own country. They are different methods; from introduce infected people, to infect soil, part atmosphere there or to export food which contain genetic code made factors which will have particular influence on people.

Next way is to introduce in country vertebrate and these in many countries done lot of damage. Larva of schistosoma haematobium is cause of suffering often many people in some with warm climate countries make them decrepit what have effect on country economy and millitary efficency. So to transport this larva with genetic improvement as will survive in more cold climate or similar way could make lot of damage too. Amphibians and their related species are many thousand so by molecular change could be adopt to more difficult conditions with similar effect for organism of vertebrate including humans by throw them to water or else where could be reproduce. Effect of schistosoma haematobium could be seen for instance in Sahara. Yet next example would be tse tse fly(glosima) of which is about 20 species. So move them from Africa on some damp areas elsewhere with genetic adaptation may done bad effect on part of population despite known cure for it. All for this have to be of course in laboratory make trials to adapt these species to new conditions. Also know ants from Amazon area which often do damage on many areas as was done similar with brasilian bees mixed with african, or locust also. All this or part move to new conditions and there adapt could do large damage for country without millitary operation. Just to move them at present is no any difficulty.

Also it is way to use on plants in particular cultivated some species symplyta or other in which with help phytopatology so in some part of year maybe apply for crops destruction if become necessary. If done properly country may lost most food reserve.

Still not enough is use possibility listening devices technology which can drop amount of people employ in spy service, yet now with big development electronic and microelectronic. Listening devices could be carry in cloth, papers, suitcase on many conference, meeting or install in other devices, furniture, wall, ceillings in many industry enterprise or millitary office, cover with paint or other material make it more difficult to find. With this that these devices should have long range of receive. As receiver can from far distance receive sound or picture and if it is in foreign country maybe install in special car or diplomatic institution. If for instance such listening device will be install in or close to research center then could pass information to near city and from there by satellite to own country with frequency change waves for security or send hide inside laser beam. So in own country in headquarters it will be receive information in few seconds. And when become necessary to make fast decision so will pass by satellite straight to diplomatic institution or directly to person which have have listening devices in other country.

It will help save save time and employment more people in transfer information and often important decisions maybe undertake in seconds from time receive information from so far away conutry. Radio waves use for communication with help with bit magnetisation maybe direct inside laser beam to send information so would be not so easy discover. All this can be done of course by many countries. In this way would be possible to install at least tens listening devices difficult to find and in many countries araund world to have information in seconds in main office of own country with use strong receivers able to catch waves from far distance when installed in commmercial outpost in different city araund world or diplomatic institutions yet in other place there. And without necessity use persons straight from these institutions, outpost pass by satellite to own country. These matters are in reach of seconds and in technology possibility. Is better to change frequency in institutions because of wire tapping. Also if these devices will be found so in other place could be replace yet more difficult to find it. Is good too use listening device production of other countries so if after will be found will be more hard to assume which country use them. These devices maybe install to radio, television, computer and electronic appliance exported to other country where they can play similar role like former one and in similar way but because of mass production more difficult to discover in which one been place. Instead of these listening devices can be in similar way install different one which activity will be not the same as former as they will be transmitters so they will send waves just that in this case it will be radioactive waves and similar which have dangerous influence on biological body but dont do material destruction. Need to be use also in case own defence if necessery.

Main source of transmission will be if need in other country so from one of few stations will be send waves by satellites or directly to small receivers, devices from where after modulation will be send in shape of radioactive rays in one or few direction on distance short and long as need for target influence. These devices by optic materials could disperse rays beam on greater range being powered by small battery or part of them send beams will power these devices after next modulation. With present progress it could be very small devices but with large strenght- even place in walls as for some rays it not make obstacle. In main point in city, stations, close to millitary base, well hide, or on ships, planes close to territory other country. This of course without large amount of them, with prudence done not to make suspission. If would be smuggle to country many these devices then in case of war it maybe more important than ballistic intercontinental missiles with nuclear weapons in dependence done such action.

In case invasion on smaller country with whole millitary power few days at least before can be switch on many these devices from more existing in attacking bigger power country in few points there and when people or army start get some sickness symptoms can be claim that was switch only part of these devices and will follow much more if invasion will continue so many more will start again.

This dont have to be true but country would dont know how many really these devices is on his territory so may be doubt if invasion will continue. These devices dont have always send rays to destroy life just could be enough to make in people prolonged sickness. If many devices will be found and destroy then need to send waves from far away for instance five next devices from other five. Then claim because still invation happen that apparatus sending waves are hide in safe place and will by itself automatic send waves ones a while to target in attacking country. It cant be only half true but country in which this happen may dont known this. So is chance that all millitary power and money spend on this may dont have that much meaning against much less use of time and money also. As it was properly done strategy connect with modern technology. If have to spend billions on conventional weapons so is better to spend much less for strategy to install these apparatus and train people as to action movement this strategy. This yet with combine before mention strategy. If country will have fear of invasion from few countries so can place these devices in many these countries and in case from own territory these waves send. Important factor is here surprise, proper ways set them and in right time so not make damage to itself. Principle devices action are known from radio transmitters where they can be place too even with higher risk. But here different kind will be use and waves range will be other also.

By modulation could be send also radioactive waves to radio, television and computer center in opponent country. In dependence from kind and strenght send waves dont have always destroy organism just develop some disease. Dont mention farther effect as panic done in country as standard waves use in communication have influence on organism by receipt of sound or light waves dont mention about strong radioactive waves. Present use electromagnetic waves send to radio, television pass by because of size by human body which is collect with atoms also as well and known elementary particles and waves. So radio waves despite different lenght get in touch with these elementary particles and waves in humans have some influence by this. By sending radioactive waves to other country have to be put attention on strenght and frequency transmitter and try with these radioactive waves disturb sound and picture in radio, television receivers in other country when unlike principles apply to computer, telephone system. Is no difficult to assume what effect could make application of waves in few countries where these waves can be transmitted(cover their frequency with already use waves) to receivers but too on war ships or millitary center and main office there. With some percentage which are spend on research and production nuclear weapons will be better develop ways of build radioactive transmiiters from own country to other or many at ones in part of second send waves, yet that action these ways dont do material damage as nuclear weapons and are more practical to use.

Is to possibility to send them early to locate there that is in opponent country receivers before desribe. So from sending them may have start result in few seconds in many countries at ones, yet on large space if necessary. Radioactive source to send such waves as plutonium etc is enough and not much here is need. But is better done cautious it as here even individual persons could do it to some range- millitary goernmment action dont mention and dose over 200-400 R would be deadly. Nuclear bombs and long range missiles are after develops this technology absolote, slow and less effective yet less safe. Also in case of attack on own country could be mention that these transmitters located in safe place and hide will in periodic intervals send these waves even many month later. Other is also use radioactive matterials. Many countries have large number of agents on territory foreign countries. If necessary they mabe direct to this kind of goals to do it. So before after receive information where are radioactive waste, most in deep shaft, so in proper safe clothes-could from such shaft part such radioactive matterial of high, average strenght take out and some of them place in these part of country where happen eartquakes or vulcano eruption. There was examples that when was bury radioactive waste to shafts so after time there happen later weak earthquakes. After remove these waste earthquakes in these place stop.

Are not known yet all activity which are happen on level known elementary particles and their action between them. But here is dependence to place such waste in their proper safe cover, amount and structure, type soil and tectonic movement there. Dont mention about throw them to vulcano without any security measures for bigger effect. Also these radioactive waste for security can be gradually send to outer space far away with cosmic going missions.

These waste in war time if need can be place close to millitary base then exploded for bigger effect, or in small dose put in water reservoir there but not in large amount to make suspition. Beside place them also in few country points to explode if will come need for this. Would be mention also about possibility which is known to send agents to and explode for damage power stations and main important for economy objects in country and they maybe explode from distance from remote control device ones material was first place there. Also from some distance is possible these target damage from laser guns which is effective method too. Next method is degradtion part of society in this case by support from other country use of drugs. To train agents in mission to help transport and delivery all kind drugs to one of few countries with caution without to fast action so eventually will do more damage to opponent country. It is like support ciggaretes smoking or spirit over drink support. But this is can be use yet better.

It is known that some infections would be pass by blood. Then is no difficult in laboratory breed some virus and add to blood. Small dose next dry and mixed with drugs which are smuggle to country so may use later to attack. There habit of drug use in some part society is spread so many will be infect in short time comparatively. As little dose dry blood with virus or bacterium transfer to organism with drugs by smoke or injection could reproduce again. To these drugs could be add cancerigenic compounds to do similar damage or similar type bacterium, virus depend from need and apply method. Later will be difficult to find if these people have been infected in other or own country. Here is enough support organisations or individuals for such mission done often by third country to deceive true source of action. Yet well train agents may do the same as well so it will find susceptible ground in country, among other in few night clubs where often abnormal sounds which is try by some claim to be a music will favour more mental degeneration. This effectively lead operation may destroy without millitary operation few countries sometime for longer period when save own and better if in moral way it stand higher if is so. But for example if are admires drug use and often abnormal sound in clubs or else, over drink, over smoking, so all this is not evidence high morality and development. Human weakness also should be to paticular degree too for own safety.

Agents use is too for blow up power stations, dams on rivers or could be smuggle ready radioactive materials to place where need. It is no pleasaure in this but effective especially if no much choice.

By throw to vulcano radioactive material they may change heat reaction, pressure to make explosion, even in geyers also. These radioactive materials is possible smuggle and secure in little package some time before expected attack foreign country on own. Few these radioactive package would be place in few country points then explosion done by own agents or by radio waves from diplomatic missions, yet even from own country. Good strategy here can do more than whole army and country defense system as could be place such weapons in two point of country then claim that will be exploded next eight like these elsewhere when it will be only two of them but other country may dont know about how many was place there. So if small country will smuggle and install four hightly and secure radioactive bombs and few next in different one if was in conflict with them and at time before expected attack done explosion part of them, claim that will later make few more explosion if attack will start so would be little doubt if will be attack. By this strategy alone may save plenty of cost and millitary operation if properly executed.

Beside less amount radioactive material or radioactive gas would be add to industrial gas in country which attack what will show fast effect even on broad scale.

Is way to use rockets with radioactive gas close to enemy border with the way straight winds, storms calculate range and direction next explode load so will go over enemy territory and his army. On present technological progress, even this way also is no need spend so much funds on weapons. Next example would be development biologicla weapons and strategy their use. Like not infectious cancer turn to infectious or ebola virus maybe smuggled and secure on country before it attack where well train agent would throw them araund army and millitary base and less dangerous on small scale seperate population so in few weeks may suffer even one third of population with chance not to point out who did such thing with correctly conduct mission when main ways to spread would be air and water. It is no moral to do it but is no moral also every other way to end life, maybe with exception own defence. In this case much cheaper than conventional and nuclear weapons with bigger devastation and after effects, plus that some of these things are known, some may come later so preparation for this can be done early.

To avoid be discover by radar, surface of missile, plane had be cover in seperate thin layer like eleoctromagnetic one yet that these layer waves have be switch in proper moment with strong magnetic waves similar in vibration and lenght to radar waves which when reach surface dont rebound back just pull back by strong magnetic waves or electromagnetic so missile or plane will be not notice. These waves can be made from generator, battery use for power in missile, plane to circulate araund them in this cover layer, few layers if need with increase magnetic power to pull waves send from radar.

This all is part matter and energy action. Matter of course first then energy which is consider as everything what is work in these condition until discovering something else. Then next motion as on this scale is possible imagine energy but without motion with different occurence still in range of work or on higher principles than motion. After these mention occurences come next quantity of physical forces. Theories are different assume and change later too so what is about just to improve them. Like here local theory about dinosaurs extinction suppose by meteor coming to earth what is no to much logical. Here could be other possibility

Is consider that DNA passing by generations is steady and will run without end but same as human average life span or other species is more or less stabilize on some amount of years so the same could be with lenght of whole species which by early division maybe settle on tens or hundred millions years for example. But not without end as is consider often and this maybe reason death rate of species by evolotion history, including dinosaurs. Such decline may last millions of years when occasionally climate condition or food change have influence yet inside change,or genes change could lead to mutation and decline species with different reason and lessen immunogical resistance, breeding decline, yet other. So lenght of life species as average their life span could be to some limit also and this maybe one of reason their decline. For last few thousand years not mention millions many species already gone, part because human activity, and still it happen. So from point hundred millions years in future one may wonder what happen that so many species gone in this time when really was no any catastrophe.

Other theory could be that if Universe was create in this part of third dimension in rotate motion like atoms, planets and have more denset mass than outside Universe space so particles from this space much smaller than neutrinos should pass this Universe and other too with next object in third dimension with zero mass against this cosmic space but not beside it with speed faster than light- so that is why is so difficult to discover them. But with science and technology development is chance for it. Even now is possible so distant before thing that is to produce tens of thousand square kilometres strong transparent or not material, roll in hunk then send to outer space then gradually link in one whole material so it would enclose earth or part of earth passing only sun rays. To make such material in large amonut is no hard to do it.

More difficult is circle araund earth but only if about time lenght no so much as technical problem. In farther future it chance to do it araund sun with material resistant to high temperature with few layers if need It is not only for far sun expansion which be in few billion years but these technologies are in range. Is way also to use them over earth against meteorites coming to push them away though they can be first cut to pieces with laser close range attach to rockets.

On small scale such material very resistant to stroke from outside could be spread by helicopters over city if in danger from missiles attack- even few layers of them to lessen missiles. Bomb, laser beams attack, also cover overs forest fire. Or so spread and resistant could be send attach to small rockets against higher amount planes and small missiles and rockect which attack just to slow them bit or push away. In far future these much stronger materials with range of link hundreds thousand kilometers yet more could cover araund moon or some planet and connect to hundreds or thousands rockets improved from these now and next could move moon or planet to new zone with power and speed bigger than known now. Beside on again lessen scale araund meteorites to cover them and direct with rockets in other direction. Next example future technology is build on moon or some planet big, transparent dome fill up with air then use including with plant cultivation. Still odd technology is to make cable araund earth set and attach to satellites on small rolles with fast spin. Such cable would be for use for fast transport object attached to it araund earth then send down by known ways.

Other example for future would be wire communication to moon if wanted as to rocket maybe attach very thin cable and fly with it to moon to attach there calcutate moon not perfect orbit. Or fly with cable araund moon then back araund earth as well. So to other planet.Yet just to satellite to attach thin cable and by electric motor power later to transport object to outer space. Next possible technology is place mirrors over earth reflecting after concentration sun rays or from artificial source to direct on few earth regions to worm them up like polar regions. By worm up little temperature in these region would be cultivate grains and fruits on bigger scale. Similar to do it is chance on moon and few solar system planets place mirror to worm them on surface and bring later oxygen and water if is there frozen, so by heat turn to water. Next example here is use during war time gas missiles which after small explosion dont kill but spread substance to weaken people for time being or put to sleep. If on whole army or city will be by surprise attack with these missiles is chance to achieve fast effect without material damage and loss of life. Yet in this case surprise attack is important with these missiles so opponent will be no ready for this kind of weapons.

This all is again describe in general usually without more specific details but in future it may come to use it and next new technologies as well for progress.

They are already innovatios to cure most disease and to travel in space with speed of light or almost any speed but typical they are left and by this hold progress instead improving with better living conditions.

1992

www.ingramcontent.com/pod-product-compliance
Lightning Source LLC
Chambersburg PA
CBHW031431210526
45464CB00005B/2147